TINPLATE AND MODERN CANMAKING TECHNOLOGY

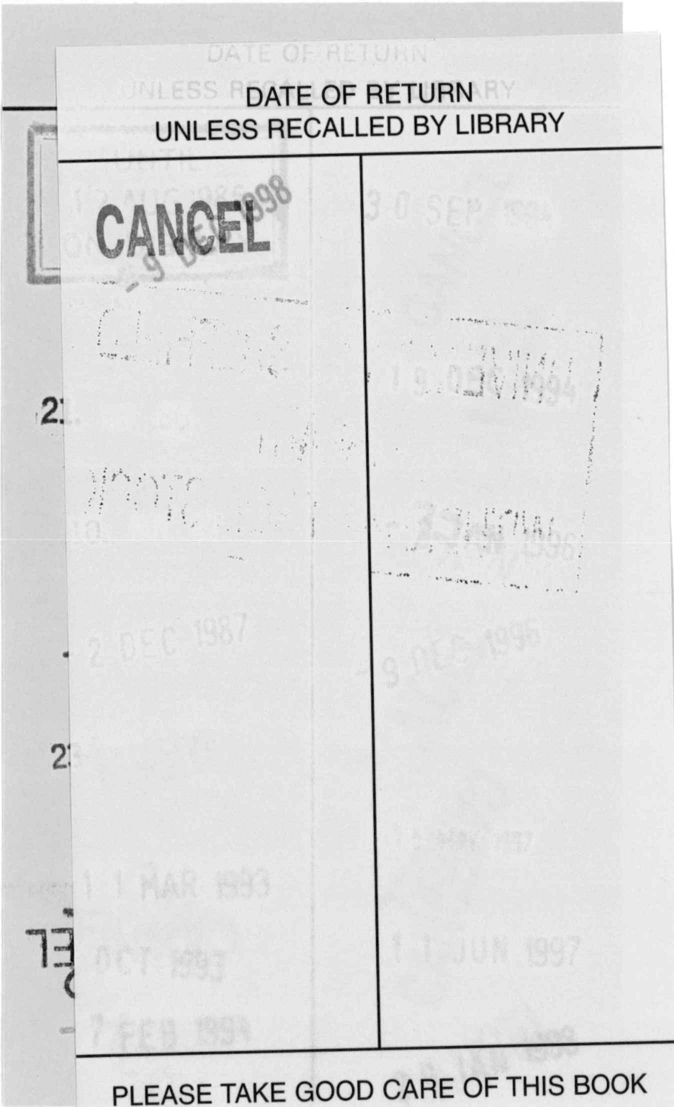

THE PERGAMON MATERIALS ENGINEERING PRACTICE SERIES

OTHER TITLES IN THE SERIES

ALLSOP & KENNEDY	Pressure Diecasting, Part 2 The Technology of the Casting and the Die
BAKER	Introduction to Aluminium Alloys
BYE	Portland Cement
DAVIES	Protection of Industrial Power Systems
DAWSON	Joining of Non Ferrous Metals
HENLEY	Anodic Oxidation of Aluminium and its Alloys
HOLLAND	Microcomputers for Process Control
LANDSDOWN	Lubrication
NEMENYI	Controlled Atmospheres for Heat Treatment
PARRISH & HARPER	Production Gas Carburising
UPTON	Pressure Diecasting, Part 1 Metal — Machines — Furnaces
WILLIAMS	Troubleshooting on Microprocessor Based Systems

TINPLATE AND MODERN CANMAKING TECHNOLOGY

E. MORGAN

Formerly Manager, Materials Development,
R & D Division, Metal Box plc

PERGAMON PRESS

OXFORD · NEW YORK · TORONTO · SYDNEY · PARIS · FRANKFURT

U.K.	Pergamon Press Ltd., Headington Hill Hall, Oxford OX3 0BW, England
U.S.A.	Pergamon Press Inc., Maxwell House, Fairview Park, Elmsford, New York 10523, U.S.A.
CANADA	Pergamon Press Canada Ltd., Suite 104, 150 Consumers Road, Willowdale, Ontario M2J 1P9, Canada
AUSTRALIA	Pergamon Press (Aust.) Pty. Ltd., P.O. Box 544, Potts Point, N.S.W. 2011, Australia
FRANCE	Pergamon Press SARL, 24 rue des Ecoles, 75240 Paris, Cedex 05, France
FEDERAL REPUBLIC OF GERMANY	Pergamon Press GmbH, Hammerweg 6, D-6242 Kronberg-Taunus, Federal Republic of Germany

First edition 1985

Library of Congress Cataloging in Publication Data

Morgan, E.
Tinplate and modern canmaking technology.
(The Pergamon materials engineering practice series)
Includes bibliographical references and index.
1. Tin plate. 2. Tin cans. I. Title. II. Series.
TS590.M67 1984 672.8′23 84-3079

British Library Cataloguing in Publication Data

Morgan, E.
Tinplate and modern canmaking
technology — (The Pergamon materials
engineering practice series)
1. Tin plate
I. Title
672.8′23 TS590
ISBN 0-08-028681-X (Hardcover)
ISBN 0-08-028680-1 (Flexicover)

Printed in Great Britain by A. Wheaton & Co., Exeter.

To Rae
for her unfailing patience

Materials Engineering Practice

FOREWORD

The title of this new series of books "Materials Engineering Practice" is well chosen since it brings to our attention that an era where science, technology and engineering condition our material standards of living, the effectiveness of practical skills in translating concepts and designs from the imagination or drawing board to commercial reality, is the ultimate test by which an industrial economy succeeds.

The economic wealth of this country is based principally upon the transformation and manipulation of *materials* through *engineering practice*. Every material, metals and their alloys and the vast range of ceramics and polymers has characteristics which requires specialist knowledge to get the best of them in practice, and this series is intended to offer a distillation of the best practices based on increasing understanding of the subtleties of material properties and behaviour and on improving experience internationally. Thus the series covers or will cover such diverse areas of practical interest as surface treatments, joining methods, process practices, inspection techniques and many other features concerned with materials engineering.

It is hoped that the reader will use this book as the base on which to develop his own excellence and perhaps his own practices as a result of his experience and that these personal developments will find their way into later editions for future readers. In past years it may well have been true that if a man made a better mousetrap the world would beat a path to his door. Today however to make a better mousetrap requires more direct communication between those who know how to make the better mousetrap and those who wish to know. Hopefully this series will make its contribution towards improving these exchanges.

MONTY FINNISTON

Preface

The advances made in tinplate and conversion technology over the last 25 years or so have been quite remarkable. Many of the current highly sophisticated processes would have been no more than visions in the thirties when research and development really started in these fields.

These have been achieved partly as a result of external threats, but are also due to a very high degree of co-operation between plate manufacturers and canmakers.

Many highly technical papers written by specialists for specialists have been published, but it has been felt that a more general account of the whole field, written primarily for technicians rather than the university student, would be of value.

Having been concerned with several aspects of these activities for many years, with close contact with tinplate manufacturers and canmakers in many parts of the world, I was persuaded to draft a book of this type.

A broad account of tinplate, particularly how its production conditions will affect its behaviour in canmaking and in subsequent use, is vital to workers in these fields; this has been included in an attempt to answer the oft asked question "What is tinplate and how does it perform?"

It is hoped that other workers in this field, production and quality control personnel and engineers also will find parts of it of value.

This is an extremely broad subject, and more than just an introduction would be impossible in one short book. As interests will vary, an attempt has been made to include numerous references which will add to the text, together with an introduction at the end of each chapter to advanced treatments of various aspects. I trust that none of the more valuable publications in this respect have been omitted.

Having "retired" a few years ago, the necessary basic preparation for the book would have been impossible without considerable help from librarians and former colleagues. The useful comments received, and search help from the Information Officers at the British Steel Corporation, Tinplate Group and Metal Box R. & D. Division have been extremely valuable. The help and guidance given by the Editor Mr D.W. Hopkins also has been much appreciated.

But of course all the opinions expressed in the book are entirely mine, and no blame is attachable to any other.

Thanks are also due to the Metal Box plc, the International Tin Research Institute and others for ready permission to publish many of the figures and photographs.

Acknowledgements

The author wishes to make the following acknowledgements for illustrations appearing in this volume:

Figures 2.2.1 and 2.2.2, Peacey and Davenport: *The Iron Blast Furnace* (Pergamon Press, Oxford, 1979).
2.2.3 and 2.4.3, US Steel: Making, Shaping and Heat Treatment of Steel (US Steel Corporation, Pittsburgh, Pa., USA, 1970).
2.4.5, 2.4.6, 2.4.7 and 2.4.8, Open University: Steelmaking Course T352 (OU, 1979).
2.5.2, Sharp: *Elements of Steelmaking Practice* (Pergamon Press, Oxford, 1966).
2.5.3, Concast Documentation Centre (AG Zürich, 1981).
2.5.5, Ingot Technology (Fuel and Metals Journal Ltd, 1973).
2.6.2, 2.6.5, 2.8.1, 2.9.3, 2.9.4, 2.10.1, 2.11.2 (a) and (b), 3.3.3, 3.3.4 and 5.1.4, Hoare: *The Technology of Tinplate* (Arnold, London 1965).
2.6.3, Datsko: *Material Properties and Manufacturing Processes* (1966).
2.6.4, 2.9.1, 2.9.2 and 5.2.2, Dieter: *Mechanical Metallurgy* (McGraw Hill, New York, USA, 1973)
2.8.2, McManus: *Iron Age* (August 1980).
2.8.3, Dunlop: *33 Magazine* (1976).
2.9.5, *Steel Times* (June 1977).
2.9.7, Price and Vaughan: ISI Symposium (The Metals Society, 1966).
2.10.2, 3.3.1 and 5.6.3, The International Tin Research Institute.
2.12.1, 2.12.2, 7.1.1, 7.1.2, 7.1.3 and 7.1.4, Gabe: *Principles of Metal Surface Treatment and Protection,* 2nd Edition (Pergamon Press, Oxford, 1978).
4.3.9, 4.5.1, 5.4.8, 6.2.1, 6.2.3 and 6.2.4, Metal Box plc.
4.4.2 (a), (b) and (c), The General Electric Co. Ltd., Hirst Research Centre, Wembley.
4.4.3, Collier *et al*: 2nd International Tinplate Conference (ITRI, London, 1980).

4.4.4 and 4.4.5, McFarlane: 2nd International Tinplate Conference (ITRI, London, 1980).
4.5.3, Schaerer: 2nd International Tinplate Conference (ITRI, London, 1980).
4.5.4, G F Norman, Metal Box plc.
4.5.5, 4.5.6, 4.5.7, (a) and (b), and 4.5.9, Sodeck: 2nd International Tinplate Conference (ITRI, London, 1980).
5.2.1, Harris: *Mechanical Working of Metals* (Pergamon Press, Oxford, 1983).
5.5.1, Fidler: 1st International Tinplate Conference (ITRI, London, 1976).
5.5.2 and 5.5.3, Jenkins: 1st International Tinplate Conference (ITRI, London, 1976).
5.6.1, International Tin Research Institute (1976).
5.6.2, *Sheet Metal Industries* (June, 1976).
5.7.1, 5.7.2 and 5.7.3, Schuler: *Metal Forming* (Goeppingen, W Germany, 1966).
6.3.1, 6.3.2 and 6.4.1, Holt: *Journal of the Oil and Colour Chemists Association* (1976).
7.3.1, Scully: *Fundamentals of Corrosion,* 2nd Edition (Pergamon Press, Oxford, 1975).
8.1.1, Thomson: 2nd International Tinplate Conference (ITRI, London, 1980).
8.2.1, 8.2.2, 8.2.4 and 8.3.1, Linley: 1st International Tinplate Conference (ITRI, London, 1976).
8.3.2, Incpen Report (1982).
8.4.1, South Yorkshire County Council.

Contents

Chapter 1

Introduction

Modern tinplate consists of a thin sheet of low carbon steel covered with a very thin coating of commercially pure tin. It possesses several important advantages:

(i) Good strength combined with excellent drawability — these arise from the grade of steel selected and the processing conditions employed in its manufacture.
(ii) Good solderability and corrosion resistance and an attractive appearance, due to the unique properties of tin.

Records show that an early version, consisting of iron hammered into thin sheets and coated with tin by dipping in the molten metal, was available in Bavaria in the fourteenth century, and that its manufacture extended across Europe over the following centuries. The first record of a successful tinplate plant in the United Kingdom was of one set up in South Wales in the early part of the eighteenth century using rolls to produce the thin sheet. By the beginning of the nineteenth century numerous small plants were in operation, again mostly in South Wales. The adoption of steel in place of iron, coupled with rolling in place of hammering, and the development of machines to replace manual labour, particularly automatic tinning machines, led to a substantial increase in output and in improved quality. Its manufacture also extended in other European countries and later in the nineteenth century rapidly in the United States. A detailed historical account is given by Hoare[(1)]* and also in the ITRI *A Guide to Tinplate* [(2)].

The major expansion in production up to the current world output of some 14 million tonnes (estimated) has been achieved by substantial improvements to manufacturing processes, culminating in sophisticated steelmaking, precision highspeed rolling mills and tincoating methods, all controlling the product to fine limits.

* Superscript numbers refer to References at ends of chapters.

Its uses have also changed considerably over the years; previously substantial proportions were used for general household articles such as plates and pots, miscellaneous boxes, etc., but in recent years the bulk has been used in the manufacture of hermetically sealed containers for packaging a wide range of food and beverages.

This dramatic change was initiated by Appert's demonstration in France in 1809 that food packed in sealed bottles and subsequently heated in boiling water (sterilised) could be preserved for a long time. In England Peter Durand obtained a patent to cover the application of this process to metal containers, and in 1812 Hall and Dankin started to produce commercially "tinned" preserved foods (the containers/products are now generally referred to as cans/ canned foods). Early containers were made laboriously by hand cutting and assembling followed by joining the two ends and body by hand soldering. Development of fast automatic canmaking and cannery equipment has resulted in rapid growth to the current very high rate of production.

Considerable quantities are also being used for aerosol containers for products ranging widely from air fresheners, fly killers and insecticides to toiletries; for containers for general products (paints, oils, polishes, etc., a wide range of closures for glass jars, and a number of engineering components.

Although the wide acceptance of tinplate is due primarily to its rather special properties, the extensive development of new can manufacturing processes and designs has resulted in a considerable increase in the number of tinplate varieties being available, not only in respect of gauge, sheet size and tincoating weight, but also in metallurgical characteristics, surface roughness and anti-corrosion (passivation) treatments of the tinned surface. In many instances special grades have been developed for specific canmaking techniques, e.g. for Drawing and Wall-ironing, Draw and Redraw and Easy-Open ends. Their development has required close cooperation between tinplate manufacturer and canmaking, and this is nowadays the norm in most countries. As will be seen later, some of the numerous variants are not in fact coated with tin.

As might be expected, the many developments over the last 50 years or so have required considerable research effort world-wide by all plate manufacturers, canmakers, other raw material suppliers and packaging research organisations, and the International Tin Research Institute (ITRI), based in England. Consequently, a vast array of technical papers and conference proceedings have been published on many of the facets involved. Many of these organisations also publish information notes which are generally available.

In particular, the ITRI, which has some eight or nine offices throughout other countries, issues numerous publications on the properties of tin and tinplate, its testing, soldering characteristics and corrosion behaviour. The International Tinplate Conference, the first of which was presented in 1976 and is scheduled to be held every 4 years, has proved to be particularly valuable in drawing together expert contributions on practically all aspects of tinplate manufacture and canmaking and their performance. Some 82 papers have been presented in the first two and the next will be held in 1984, with the published proceedings available from the ITRI in London.[2]

The availability of tinplate containers capable of maintaining a very wide variety of foods and beverages in a sterile state, under conditions ranging from arctic to equatorial for long periods, has made it possible to transport essential and luxury foods over very long distances and to make them available for consumption at almost any time of the year. Expeditions can be of longer range and the relief of populations after natural or other disasters can be more effective and rapid. General products, such as paints and oils, can be widely distributed in convenient amounts. The combination of mechanical properties and formability exhibited by tinplate have made it the prime material for fabrication of the pressurised containers necessary for the aerosols which must be stored and dispensed with convenience and safety.

The demand for effective packaging is always under severe cost pressure and from competitive materials and methods, and there will always be a strong demand for improved quality and design at minimum increase in cost. Development efforts will therefore need to be continued, probably on an increasing scale.

As a better knowledge of the manufacturing processes used will be of real help to those concerned with developing new forming processes and container design, Chapter 2 aims at describing the factors in steelmaking, hot and cold rolling, and electroplating which have a significant effect on the forming characteristics and service performance of the material. This area probably demands the greatest co-operation between all, from steelmakers to the consumers of canned goods.

As the properties of tinplate and other container materials inevitably vary to a degree, adequate inspection and testing are vital; most consumers will carry out this task thoroughly to complement the quality control procedures adopted by the primary manufacturers. The procedures are normally slanted to the particular requirements of the process involved; they are described in Chapter 3.

Can manufacture has been divided into two areas, the longer established in Chapter 4, and the more recent advanced technologies, still under development, in Chapter 5. In describing these, attempts have been made to highlight the more important plate characteristics.

In a similar manner, Chapter 6 deals broadly with the principles of lacquering and printing, and Chapter 7 with internal and external corrosion behaviour. Later chapters give an account of (a) extensive waste recovery methods generated over the last 10 years or so to avoid the considerable loss of primary resources and of energy in consigning used cans as refuse, and (b) legislation for improved packaging safety and reduced pollution hazards. As a deal of the substantial development undertaken over the last 50 years will continue in the future, the final chapter presents an attempt to look into the future, highlighting the improvements desired and how they are likely to be achieved.

REFERENCES

1. W.E. Hoare, E.S. Hedges and B.T.K. Barry, *The Technology of Tinplate*, Edward Arnold, London, 1965.
2. *A Guide to Tinplate*, International Tin Research Institute, London, 1983.

Further information can be obtained from:

American Iron & Steel Institute, Washington D.C. 20036, USA (1000 16th Street, N.W.)

Can Manufacturers' Institute, Washington D.C. 20036, USA (1625 Massachusetts Avenue)

Metal Packaging Manufacturers Association, Windsor, Berkshire SL4 1BN, England (3-9 Sheet Street)

International Tin Research Institute, Perivale, Greenford, Middlesex, England (and information offices in Australia, Belgium, Brazil, Germany, Holland, Italy, Japan and the U.S.A.)

National Canners Association, Washington D.C. 20036, USA (1133 20th Street, N.W.)

The Campden Food Preservation Research Association, Chipping Campden, Glos., England

and from most major steel and tinplate manufacturers and canmakers.

Chapter 2

Tinplate Manufacture

2.1 INTRODUCTION

Tinplate is essentially low carbon steel between 0.15 and 0.49 mm thick coated with between 5.6 and 22.4 grammes of tin per square metre — uncoated steel sheet and chromium/chromium oxide coated steel are also available.

Continuous demand for improved quality and more economic production have led to the development of highly sophisticated manufacturing techniques. The essential aim of this chapter is to present an outline of the manufacturing methods used and their effects on metallurgical and mechanical characteristics, in the belief that it will be of help in the development of improved and new canmaking processes.

Manufacturing methods will be found to vary appreciably, but it is probably true to say that tinplate and its related products from different suppliers are now more uniform in their characteristics than ever before.

Figure 2.1 presents an outline of the sequence of operations.

2.2 MANUFACTURE OF PIG IRON

Commercial extraction of iron from its ores is predominantly carried out in blast furnaces which are basically tall "vertical shaft" furnaces in excess of 30 m high, having means for feeding the charge materials through an opening in the top and for collecting the resulting liquid iron and slag at the base as well as gas emitted at the top. The basic design and construction of a typical blast furnace is shown in Fig. 2.2.1, and its auxiliary equipment in Fig. 2.2.2. The charge will consist of a mixture of iron ores, solid fuel (coke), and fluxes (usually limestone and dolomite). This is fed almost continuously into the furnace, and a blast of hot air is blown continuously through tuyeres located above the hearth. The furnace illustrated in the figures is the Fukuyama No. 5 in Japan, typical of modern blast furnaces.

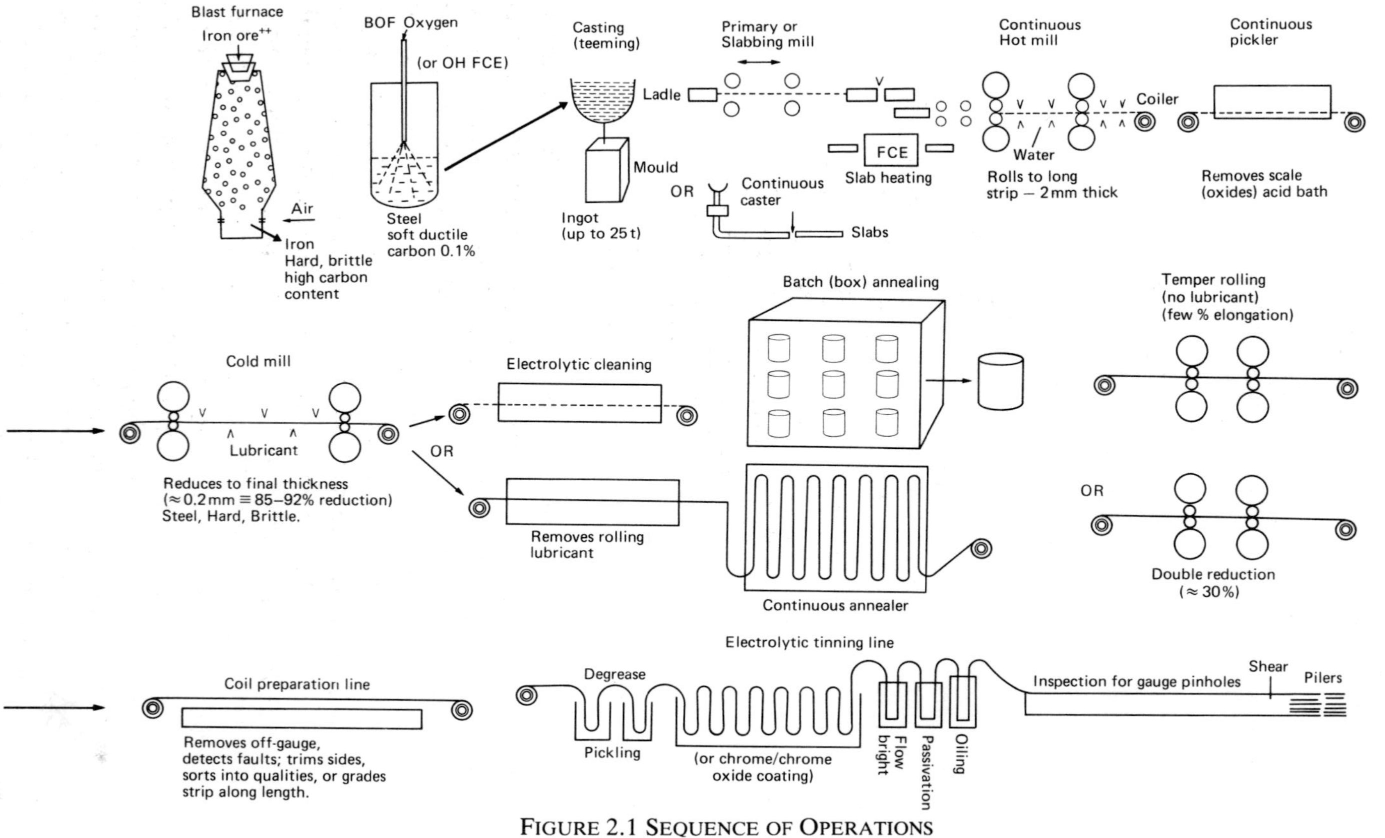

FIGURE 2.1 SEQUENCE OF OPERATIONS

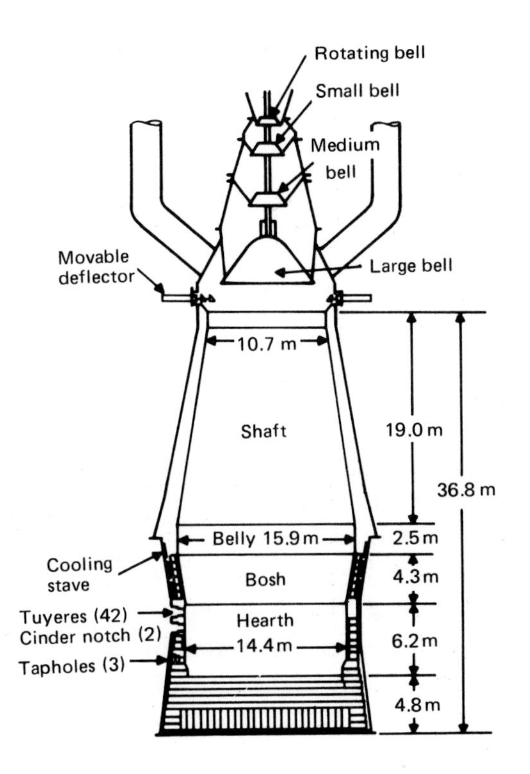

FIGURE 2.2.1 A MODERN BLAST FURNACE

The iron ores are generally haematite (Fe_2O_3) with some magnetite (Fe_3O_4), and are prepared as pellets 1-2 cm diameter made from finely ground beneficiated ore, as 1-3 cm lumps of sinter from fine ore and/or 1-5 cm screened ore. In addition, quantities of mill scale and other iron-bearing materials accruing within the integrated plant are added. Coke is included to provide the reducing gas and heat needed for reduction of the ore and smelting; the coal used to produce this coke is carefully selected to have not more than 0.5-1.0% sulphur, in addition to the particular physical characteristics needed for the operation. The CaO and MgO provided by the flux materials combine with the silica and alumina impurities in the ore and coke to produce a fluid slag at the temperature obtaining within the furnace; they also assist in the removal of sulphur. The air blown through the tuyeres is preheated up to 1300°C, and can be enriched with oxygen to increase reaction rate and it may also be mixed with gaseous, liquid or fine solid hydrocarbons, such as tar or fuel oil, natural gas or powdered coal, to provide additional reducing gas (CO and H_2).

The primary products — liquid iron and slag — are tapped frequently from the furnace — almost continuously in the large modern furnaces — through several tap holes near the base of the furnace. Liquid pig iron will normally have the following composition:

Carbon	3.5 – 5.0%	Silicon	0.3 – 1.0%
Manganese*	up to 2.5%	Phosphorus*	up to 1.0%
Sulphur	up to 0.08% but generally lower		

* Mn and P dependent on ore.

Although the iron composition is not critical, as it is smelted in the next stage — steelmaking — it must be related to the refining method being used and be consistent, so that the desired final steel composition is efficiently and economically achieved. Its composition is controlled by adjustment of the slag composition and the temperature in the lower half of the furnace.

Internal operation of a well-controlled blast furnace is very stable for many reasons, especially preheating of the air blown through at pressures up to up 3 atm. In fact, routine control of many factors is employed, e.g. temperature checks on the liquid iron and slag, on the blast prior to entry, rates of flow, and temperatures of the gases through the tuyeres and through the burden to the exit point at the top of the furnace; flow rate and temperature of the cooling water; and the composition of the iron and slag. All of these data are handled in a programmed computer operation, giving complete control over the whole operation of the furnace and its auxiliaries. After tapping, the liquid iron is run into large insulated vessels fitted on rail cars for transport to the steelmaking plant.

The chemical reactions taking place within the blast furnace are complex, and will depend on many factors; basically, the preheated air reacts with the carbon (coke) to produce carbon dioxide with a substantial amount of heat:

$$C + O_2 \rightarrow CO_2$$

The CO_2 is reduced to carbon monoxide by an excess of coke and the water vapour present is similarly reduced to CO and hydrogen. These powerful reducing agents, will reduce most of the iron oxides to metallic iron, the balance being reduced by solid carbon in the hearth.

$$CO_2 + C \rightarrow 2CO$$
$$H_2O + C \rightarrow CO + H_2$$

$$Fe_xO_y + (y)CO \rightarrow (x)Fe + (y)CO_2$$
$$Fe_xO_y + (y)H_2 \rightarrow (x)Fe + (y)H_2O$$

The temperatures obtaining in the bosh (≈1800°C) are high enough to melt the iron (m.p.≈1200°C) and the slag (m.p.≈1300°C); the iron in the hearth is saturated with carbon at the prevailing temperature and there are significant amounts of manganese, silicon, phosphorus and sulphur. The slag will absorb most of the sulphur introduced by the coke, but all of the phosphorus in the ore will pass to the iron. Gangue in the ore will react with added fluxes to form slag, which will retain unreduced oxides and impurity compounds, but will be low in iron because of the reducing conditions in the hearth.

Modern blast furnaces produce molten iron of near constant composition at high rates; they have hearth diameters approaching 15 m and are operated continuously; in fact a "campaign" can last as long as 8 years before the furnace has to be shut down for major repair (minor repairs can be carried out whilst it is in operation). The largest are capable of producing more than 14000 tonnes of iron per 24 hours. The current scale of operations is illustrated by Table 2.1 below, giving quantities per tonne of iron produced.

TABLE 2.1 QUANTITIES OF IMPUT AND OUTPUT MATERIALS PER TONNE OF IRON PRODUCED

Iron ores and other iron-bearing materials	1.6 – 1.7 tonnes
Coke	up to 0.5 tonnes
Lime-bearing fluxes	about 0.2 tonnes
Air	more than 1000Nm^3 at 1 atm plus 30 Nm^3 of oxygen
Oil or equivalent	40 – 120 kg
Added water in the blast	10 – 35 kg
Slag produced	200 – 400 kg

The use of lower grade ores will increase these figures considerably.

Other by-products will be the gas drawn from the top of the furnace, carrying some iron-containing dust, and having a calorific value of about 4000 kJ/Nm^3, equivalent to about one-tenth that of natural gas; its approximate composition is 23% CO and 3% H_2, with 22% of CO_2 and balance nitrogen by volume.

To carry out an operation of this kind requires large-scale back-up services and equipment; this is illustrated in outline in Fig. 2.2.2. They range from preparation, handling and weighing of all solid raw materials (mixtures of ores, coke and limestone), preheating and blowing of large volumes of air, collection and cleaning of gases emitted (to avoid fouling of piping and burner jets on subsequent use as fuel), and means for tapping and removing liquid iron and slag. Coke is produced from suitable coal in coke ovens, usually on an adjacent site. The air is preheated in large recuperative stoves heated by a mixture of cleaned coke oven and blast furnace gas. Substantial amounts of cooling water are also required.

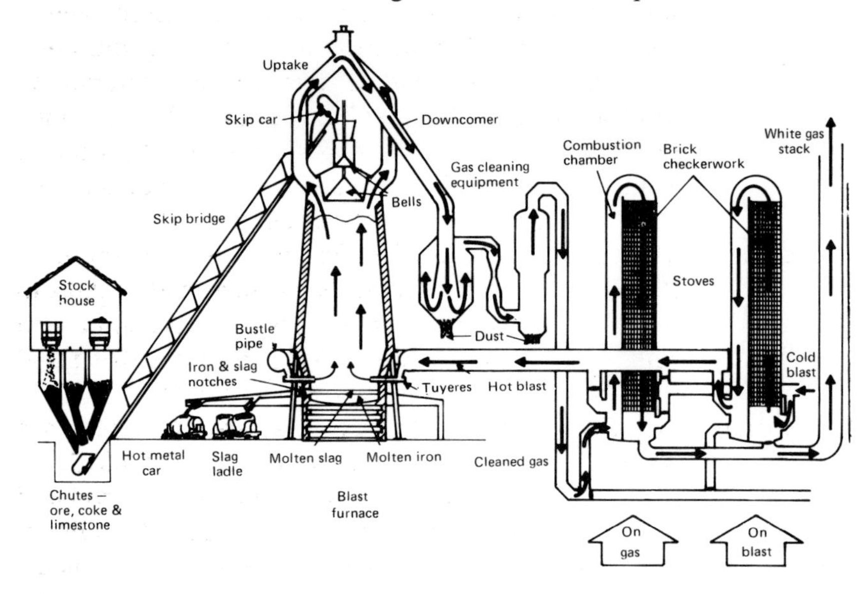

FIGURE 2.2.2 SCHEMATIC CROSS-SECTION OF A BLAST FURNACE PLANT SHOWING MATERIALS-HANDLING, CHARGING, TAPPING, GAS-HANDLING AND HOT-BLAST EQUIPMENT

To ensure the greatest reduction efficiency practicable, thereby providing greatest throughput and minimum usage of fuels, efficient distribution of gases through the charge is aimed for; this has been improved in recent years by employing an optimum size range of iron ore materials and coke.

Continuous intensive effort has been devoted to improving the productivity of the blast furnace over the last 25 years or so, and substantially higher outputs are now being achieved. Probably the greatest factor in raising reduction efficiency has been the use of iron-bearing materials of more uniform size, coupled with more effective distribution of the charge constituents within the furnace; these have given more uniformly distributed flow of gases through the charge, and more direct access to the charge burden. The improved methods of ore beneficiation are described below.

As mentioned earlier, a substantial increase in air blast temperature, made possible by the availability of more resistant refractories and better cooling techniques, have also made valuable contributions to improving efficiency.

Agglomeration processes are aimed at ensuring that up to 90% of the total ore content is within a narrow size range, usually 1-5 cm. Four methods are used:

Sintering. A mixture of 90-95% ore fines (−¼in) and 5% coke, sometimes with added limestone, is passed over a continuous moving hearth in a sintering machine. After ignition, the mass is caused to agglomerate by localised fusion, the heat being provided by combustion of the coke in air drawn through the charge. Temperatures of 1300-1400°C are reached, and the product is broken into porous lumps.

Pelletising. The ore which is too fine for sintering (−40#), or if there is no sinter machine, is crushed and ground to −350# and formed into pellets by moistening and rolling in a drum or disc. The pellets are dried and then hardened by heating to about 1100°C.

Nodulising. Ore fines and blast furnace dust are processed in a rotary kiln at high temperature (1100-1200°C) to form nodules — this process is difficult to control and is regarded as being more expensive than the others.

Briquetting. Briquettes are formed hot without the use of a binder, temperatures of the order of 1000°C being normal; fuel costs are not unduly high, but wear on briquetting rolls is costly.

Illustrated in Fig. 2.2.3.

The treatment of ores, a fascinating but complex subject, has received extensive attention over many years. Initially, high-grade ores available in the vicinity were used, but as they became exhausted, attention had to be given to the improvement of lower grade ores, involving various methods of beneficiation, including those described above. Concurrently with these developments large deposits of high-grade iron ore were discovered in many countries, particularly Canada, South America, South Africa, Australia, etc., but frequently they were located in remote areas. To economise on transport costs, mineral processing and agglomeration were carried out locally. To assist commercial exploitation, large ore carriers — with capacities of up to 500 000 tonnes — were also designed, together with appropriate mass handling equipment. The large-scale developments called for very high capital investment, coupled with problems of large-scale logistic control. Improved methods of processing iron ores received impetus from the growing need for economy in fuel costs. The siting of modern ironmaking technology adjacent to coastal areas was a further logical step, including completely integrated steel plants; this was at the expense of the traditional inland steelmaking units. Attempts have been made to apply the new technology to the smaller inland plants, but these in general have not been successful. Thus the overall trend is to plan massive coastal plants accessible to iron ore fields, using large-scale sea transport.

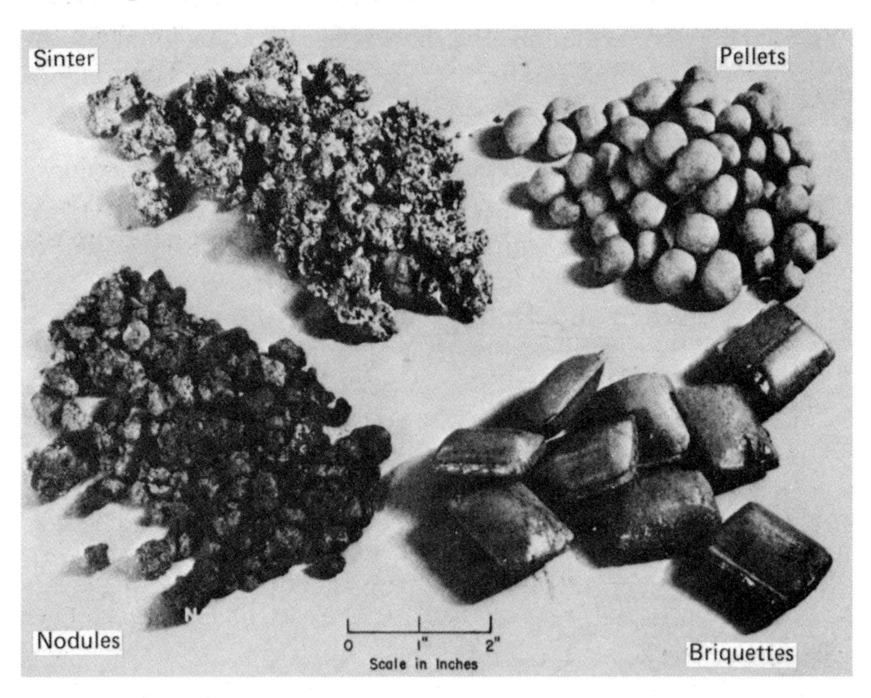

FIGURE 2.2.3 FORMS OF IRON ORE AGGREGATE FOR BLAST-FURNACE FEED

The large modern blast furnace provides an economic and efficient process, but as outputs approach 5 million tonnes per annum from each, assured large markets are required, coupled with the need for large amounts of capital. These conditions are met in only a few highly industrialised areas, and in consequence there is a trend towards designing modern efficient furnaces with annual outputs rather less than 10 000 tonnes per 24 hours. But in some less industrialised areas, the so-called "emerging nations", even these requirements cannot be met, resulting in considerable interest being taken in the possibilities of "direct reduction" processes. These techniques are described in the next section.

2.3 DIRECT REDUCTION PROCESSES

Extensive efforts have been made for many years to develop a fully acceptable process for direct reduction (DR) of iron ore to metallic iron. This has been encouraged by the heavy capital cost of the blast furnace route, coupled with its high production rate. This process will be more appropriate for many less developed areas, especially those having iron ore mines adjacent to deposits of coal which are inadequate in quality for coking purposes. Adoption of this technique has been widespread over the last decade, especially in South American countries.

The principle of direct reduction is relatively simple; dry, beneficiated oxides of iron are mixed and heated with appropriate amounts of a reducing material — any significant residues are likely to affect the product properties. A typical specification of a suitable ore or concentrate will be:

Total iron content	not less than 66%
Gangue content	not more than 4%
Phosphorus	0.03% max
Sulphur	0.02% max

Pelletised ore will range in size from 6 to 16 mm, and lump ore is screened to 6 to 40 mm; when coal is used as the reductant, it will be prepared to a similar size range to that for lump ore. Although less critical in quality than coking coal, it must have a low sulphur content.

Temperatures higher than 1000°C are used, as otherwise the rate of reduction is unacceptably low; but close control is required to avoid significant fusion and excessive pick-up of carbon.

For many years hydrocarbon-based reductants were preferred, but as a result of increasing world oil prices many attempts have been made to improve the process based on coal. Increasing oil prices have had a similar effect on the number of direct reduction plants planned and operating; the rate of growth through the seventies appears now to be slowing down appreciably. Detailed surveys have been carried out of installations employing the major DR processes — Midrex (American) and HYL (Mexican) — and several others, over the last decade and planned for the mid-eighties. The *Metal Bulletin*[1] has published a survey of the 1981 capacity and production of direct reduction iron (DRI), covering all significant processes and countries, prepared by the Midrex Corporation of America. The *Fourth Bibliographical Survey of Methods of Direct Reduction of Iron*, including a study of sponge iron sources, and the availability of ore and reducing agents in the European Coal & Steel Community, was published in 1976 (original in German), and an English version can be obtained from the Metals Society.[2] In connection with the effects of increasing oil prices on closures and cancelled projects, a note on the British Steel Corporation plant at Hunterston, completed in 1979 but never operated, is given in the *Metal Bulletin*.[3] With regard to the effects of prices of raw materials on commercial viability, the article "Direct reduction — will coal revive the flagging DR" in *Metal Bulletin Monthly*[4] is particularly informative. As part of the effort towards improving the coal-based process, the Midrex new Electrothermal Direct Reduction Process has been developed.[5]

The product has generally a minimum iron content of 92%, and a carbon range of 1.0 – 1.5%, but this is adjustable. It can be fed into steelmaking systems as part of the blast furnace charge, but its main use is in electric furnace practice, and it is therefore a competitor to steel scrap. Latterly the price of scrap has decreased appreciably, to a lower level than the cost of producing DRI, a major factor in its reduced output. The future of this technique is doubtful, but it will depend greatly on technical developments; most authorities believe that it will grow in the lesser developed areas, in view of its lower capital cost, coupled with lower outputs, especially in areas where steel scrap is scarce, and indigenous deposits of iron ore and suitable coal are available. Some favour the export of DRI rather than beneficiated ore as a more attractive proposition. A good code of shipping practice is required as the material can be potentially explosive. A recent update of efforts to develop improved DR processes based on coal and plasma heating is given by Stephenson.[6]

This technique has also been examined as a means of producing iron powder, particularly for conversion to sheet metal, or semi-finished components. Several organisations have examined this possible process, and in particular the American Can Company have combined with the Mineral Science Division of UOP Inc., following the latter's development of a process for producing high quality powder from iron ore, which was suitable for powder-to-sheet rolling, to develop jointly an integrated mini-mill based on their respective pure iron powder production and strip rolling techniques. Details of the powder production and powder rolling projects are given in *Metal Powder Review* of October 1981.[7]

2.4 STEELMAKING

The pig iron produced in the blast furnace contains, as given in section 2.2, metalloid contents far too high for steel sheet, and they must be substantially reduced in the steelmaking stage. The desired composition for tinplate will usually lie withing the following range:

Carbon	0.03-0.12%	Manganese	0.20-0.60%
Phosphorus	up to 0.15%	(above 0.04% for re-phosphorised steel only)	
Sulphur	up to 0.05%		
Copper	up to 0.20%	Silicon	less than 0.10%
		Nickel	up to 0.20%

The above covers virtually the full range of sheet steels for all purposes and tinplate base specifications are within a much narrower band. For example, phosphorus will not usually exceed 0.02%, and to ensure good ductility the carbon content will be within the lower half of the above range.

Many other elements will also be present in the steel in minor amounts, including copper, nickel, tin, chromium and nitrogen. Copper, tin, chromium and nickel will be present because of the use of contaminated scrap and the nitrogen content is determined primarily by the refining process used. Many of these elements are not removable by steelmaking processes and in view of their deleterious effects on the physical and chemical properties of the final sheet all scrap should be carefully selected and sorted. A more detailed discussion of the effects of various elements on mechanical properties and corrosion behaviour, together with the compositions used for various temper grades, is given in Chapter 3.

Steelmaking processes can be classified according to the manner in which the oxidation of impurities is effected, or the type of furnace used. Gas oxidation can be carried out in a converter, or iron oxide can be used in the slag of the reverberatory open-hearth furnace, using gas or oil as fuel. Electric arc furnaces of up to 150 tonnes capacity can operate with injected oxygen, or with slag to which iron oxide has been added, but this method of heating is not used to produce steel for tinplate to any extent. In all cases, the lining can be either acid (siliceous) or basic (dolomite/magnesite) refractory, depending on the composition of the iron used and the product desired. The balance between these processes has changed substantially in the last 30 years in that the open hearth furnace has virtually disappeared, being replaced by the gas oxidation or pneumatic processes derived initially from the Bessemer process, but modified drastically by the LD process of about 1950 and the variations introduced since that date. The utilisation of electric arc heating has been extended by adoption of oxygen injection and magnetic stirring. In all processes the furnace linings have long since been made completely basic, and commercially pure oxygen is the most widely used oxidant. The most widely used furnaces are described as Basic Oxygen Furnaces (BOF's) operating the Basic Oxygen Process (BOP) (Fig. 2.4.1).

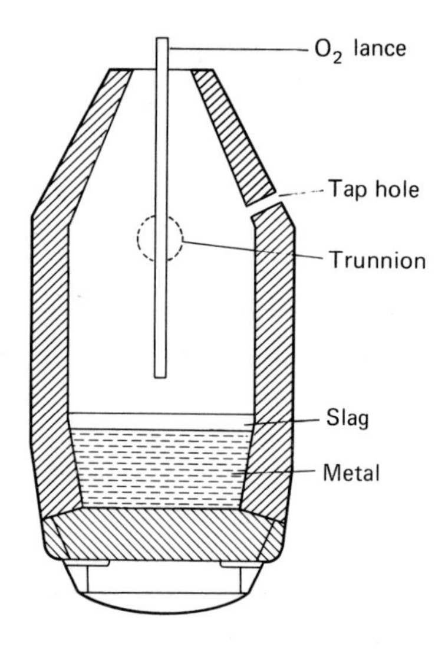

FIGURE 2.4.1 SCHEMATIC SECTION OF BOF

Some low carbon steel is still made in the Siemens – Martin open hearth furnace (Fig. 2.4.2). This is a fuel-fired reverberatory unit with regenerators, the original design dating from 1863. Refining is effected by contact between liquid metal and a lime/iron oxide slag, assisted in recent years by oxygen lances through doors and later the roof. The maximum capacity is 600 tonnes, with most furnaces in the range of 150-250 tonnes. There is an advantage in that the charge can contain up to 55% of scrap, but the refining cycle is 4-10 hours and the cost of fuel, labour and refractories, particularly since 1973, has become so high that this slow and expensive process has been displaced. Whereas usage of the BOP increased by 100% between 1965 and 1980, open hearth steel production in 1980 was only 15% of that in 1965.

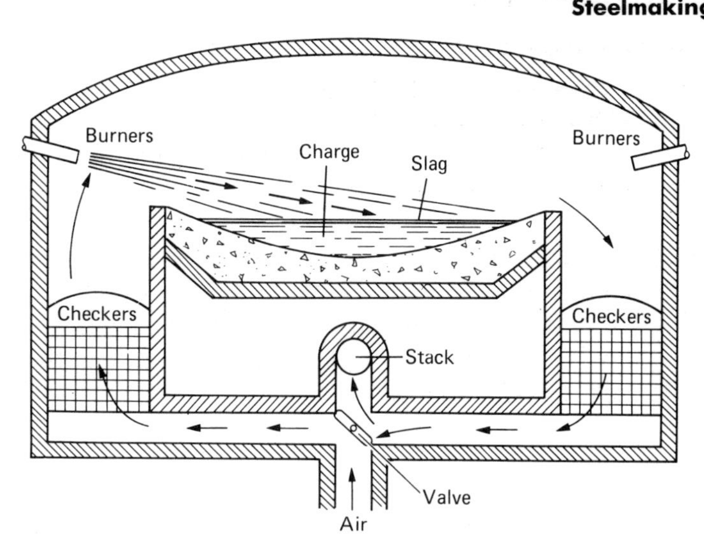

FIGURE 2.4.2 BASIC DESIGN OF AN OPEN-HEARTH FURNACE

2.4.1 The basic oxygen process (BOP)

This process is carried out in a cylindrical vessel having a hemispherical lower portion and a conical top section, the general proportions tending in recent years to the diameter being roughly equal to the height of the centre section. The whole can be tilted about a horizontal axis and there is a tap hole in the top section (Fig. 2.4.1). Most production models have a capacity of 200-300 tonnes of liquid steel and oxidation is effected by a jet of oxygen which is blown vertically down on to the surface from a suspended water-cooled copper lance, the height of which can be adjusted. A hood and cooling arrangements are provided for collection of the mainly CO gas evolved during blowing. Ladles and cranes are available for feeding, servicing and removing input and output materials. The furnace is charged with 85-95% liquid pig iron and 10-15% scrap, together with about 5% by weight of lime. The lance is lowered and the supersonic jet of oxygen is blown into the metal. This oxidises Fe to FeO and this, in turn, oxidises the Si, Mn and P and forms a slag with the lime and the SiO_2, MnO and P_2O_5 so formed with the generation of considerable quantities of heat. S is transferred to the slag as CaS and MnS and C is converted to CO. This gas causes the metal bath to "boil" and results in the formation of a gas/slag/metal emulsion which may overflow the vessel and within which reaction is extremely rapid. The lance may be raised to a higher blowing position toward the end of the blow, so as to impact, but not penetrate, the slag layer. Efficient refining of higher P pig iron is achieved by either injecting lime powder with the oxygen or pouring off the first slag and making a second one with fresh lime, leaving it in the vessel when the metal is tapped, to become the first slag of the next heat. Blowing time is about 15 minutes, the end being signalled by the cessation of CO emission and checked by analysis for C, Si, S, P and Mn. Tap-to-tap time is 40-50 minutes. Fig. 2.4.3 indicates the order and rate of removal of the elements listed.

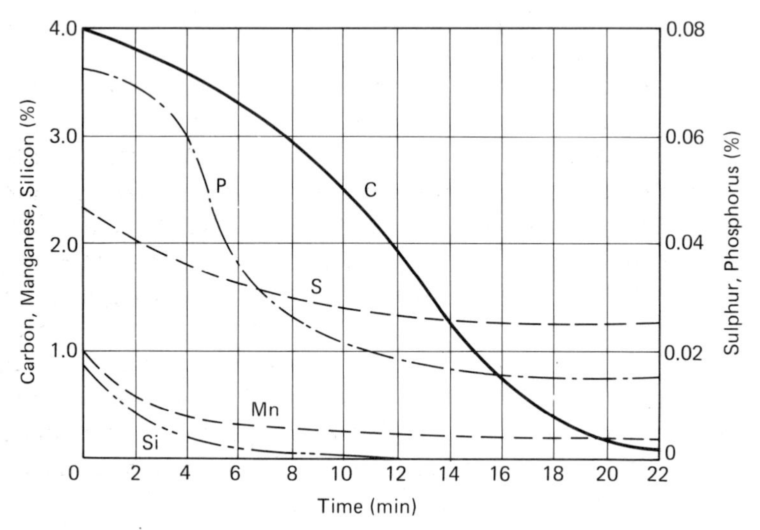

FIGURE 2.4.3 REDUCTION OF MAIN ELEMENTS IN TOP BLOWN BOF

This method can be dated from the developments at Linz and Donawitz in about 1950, and since then there have been many attempts to increase the speed and efficiency, particularly by combining top, side and bottom blowing to achieve more rapid and

Types of BOF and comparative features

Combination blowing

LD | LD-KG | K-BOP | Q-BOP

Ar or N_2 | O_2 + CaO | O_2 + CaO

Bottom blowing gas volume	None	→	Large
Agitation	Weak	→	Strong
FeO in slag	High	←	Low
Slopping	Large	←	None
Yield ratio	Low	←	High
Dephoshorization of high carbon steel	Suitable	←	Unsuitable
Dephosphorization of low carbon steel	Unsuitable	→	Suitable

FIGURE 2.4.4 BOF VARIATIONS

complete impurity removal. Rapid bottom and tuyere wear presented a serious problem, and this was ultimately dealt with by incorporating concentric tuyere nozzles through which a stream of shielding hydrocarbon gas or liquid can be passed. Argon, propane, oil and nitrogen have been used and in one case, LD-KG, inert gas stirring from underneath is combined with top blowing with oxygen. These variants have been entitled AOD, Q-BOP, STB, etc., and some are illustrated in Fig. 2.4.4.

The optimum process will depend on many factors; type of existing plant, cost of modifications and the relative economics of iron and scrap available in the area. A "green field" site will attract the most efficient modern process and an informative review of these developments is given by McManus.[8] They have resulted in the construction of highly automated plants in which charge weight and composition are linked by computer with blowing conditions and slag control to produce more uniform and cleaner steel. In addition to the lower contents of S, P and non-metallic inclusions, there is less nitrogen and so a lower liability to age-hardening. Typical nitrogen contents of BOF steel compared to the product of air-blowing is a Bessemer converter and production in an open hearth furnace are given below:

Process	N_2content
Bessemer (air)	0.010 – 0.020%
Open hearth	0.004 – 0.006%
BOF	0.001 – 0.002%

The processes which are capable of utilising high proportions of scrap are economically attractive as long as it is available at reasonable cost, but this increases the need for careful checking of composition. The problem is aggravated by the adoption of continuous casting, since this diminishes the supply of in-plant scrap, the composition of which is generally satisfactory. All of the problems of final composition and scrap procurement are specially important when making steel for deep drawing, as in the manufacture of two-piece cans.

Successful production of drawn and well-ironed (DWI) and drawn and redrawn (DRD) can bodies demands particularly clean steel, i.e. with a very low content of non-metallic inclusions (NMI). These are particles or solidified globules of oxides, sulphides, refractory and/or slag which have been entrapped in the metal. They reduce ductility and corrosion resistance and, in extreme cases, can cause fracture and give rise to pinholes. Improved steelmaking control has involved close attention to deoxidation, in order to produce the maximum of liquid oxide which can agglomerate and rise into the slag, rather than remain as very small particles which would require several hours to separate. This is especially important when the steel is cast into long slabs by continuous casting, instead of relatively short ingots.

BISRA and BSC/BISPA publications[9,10] provide useful reviews of the generation and incidence of NMI. The means adopted for ensuring that the required standard is met will vary according to plant conditions and the steelmaking system in use. Wiltshire has described in detail[11] the method adopted by the Australian Iron and Steel Pty. Ltd.

Many improvements to steelmaking processes have developed along the lines of second stage treatment of the liquid steel; these have been termed Secondary Steelmaking Processes. Almost all have a beneficial effect in reducing NMI contents, some giving marked improvement. The aims of these processes are to provide better control over steel composition, greater uniformity of temperature throughout the melt, and more effective removal of undesirable impurities from the molten metal; generally they also increase output.

Most are based on the use of reduced pressure. It is perhaps paradoxical that the benefits resulting from using vacuum treatment were described by Aitken in the *Journal of the Iron and Steel Institute* in 1886; this principle could not be applied, however, until the development of efficient steam ejectors in the early 1950s provided a means of producing high vacua at low cost, thereby allowing vacuum degassing of tonnage steel to be carried out commercially. This treatment is particularly useful for removing gaseous hydrogen, prior to applications where hydrogen embrittlement can be a problem. A general review of these valuable secondary processes is given by Houseman.[12]

Methods of applying vacuum degassing to liquid steel can be grouped into four:

(i) Stream degassing, which includes ladle-mould degassing, ladle-to-ladle degassing, tap degassing; Fig. 2.4.5 shows stream degassing, usually carried out in a vacuum tank containing the ingot mould.

(ii) Ladle degassing; Fig. 2.4.6 shows the basic simplicity of the method, and it is usually the cheapest. It is preferred for fully-killed steels (see under ingot casting below).

(iii) The DH (Dortmund-Horder) degassing technique is illustrated in Fig. 2.4.7. Metal circulation is achieved by moving the degassing vessel vertically, so that 10-15% of the metal in the ladle at a time, is lifted into the degassing vessel; and subjected to the vacuum for up to 50 cycles; periods of up to 20 minutes are necessary to ensure adequate degassing. As the pumping action induces considerable turbulence, which will persist after degassing is complete, alloy additions as required are made from overhead bins, providing a close control over the composition of the final melt.

(iv) The RH (Rheinstahl-Heraus) process is illustrated in Fig. 2.4.8. It is similar to the DH method, but a smaller vessel is capable of treating a similar given volume of steel, and can process a wide range of heat sizes. Circulation of liquid metal is initiated by reducing the density in the upper leg by the introduction of small amounts of argon. Cycling is continued until each unit volume of steel has passed through the degassing chamber typically 4-6 times.

As these processes require treatment times of up to 30 minutes, together with a high degree of vacuum and so a higher steel temperature, Nippon Steel in Japan have improved upon the RH process in several ways; these aim to degas the steel more effectively, and also to meet the high standard of cleanness, but using an appreciably shorter treatment time and a reduction in the superheat of 50-80°C:

(a) RH – CAS (composition adjustment by sealed argon bubbling) also aims at a saving in ferroalloy consumption. The CAS tube is immersed into the molten steel contained in a ladle, after "blowing aside" the slag layer, and the alloys are added to the steel under a non-oxidising atmosphere. Ten minutes treatment time is adequate.

(b) RH – Light treatment was also developed in 1975. The vacuum level is reduced as decarburisation proceeds, giving low energy consumption. Free oxygen levels are appreciably lower than in the conventional process, and lower consumption of ferroalloys results.

(c) RH – OB Light treatment is a further modification in which oxygen is blown into the RH vessel at an appropriate state; even more effective decarburisation is thus achieved.

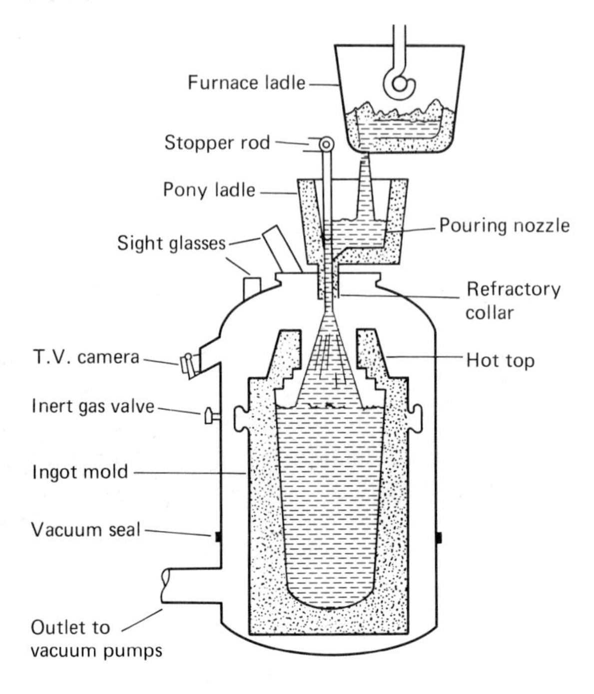

FIGURE 2.4.5 STREAM DEGASSING

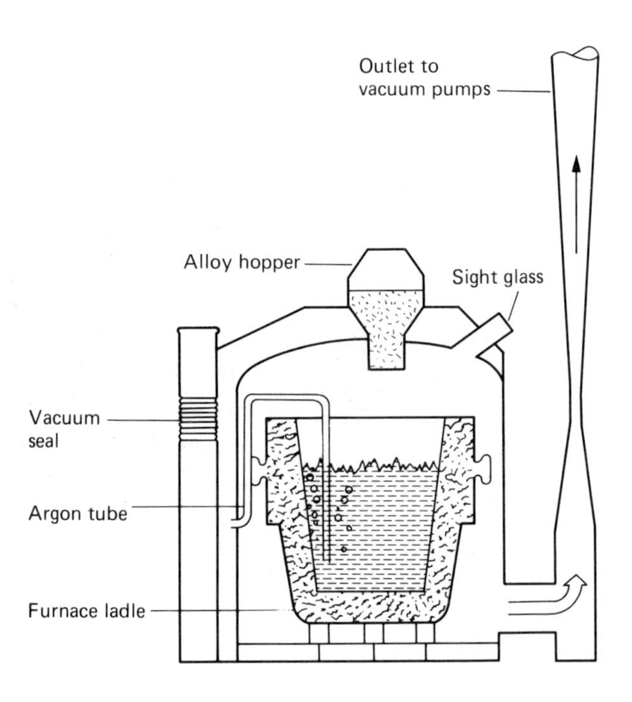

FIGURE 2.4.6 LADLE DEGASSING

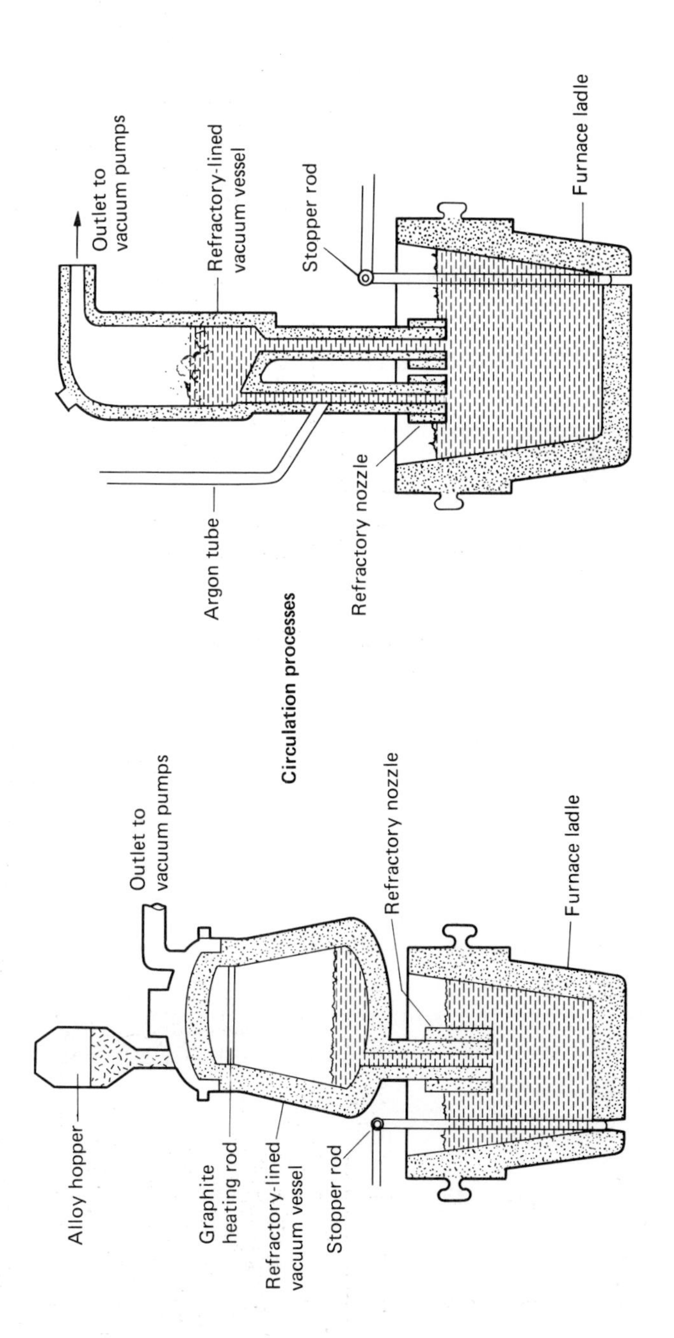

FIGURE 2.4.7 D-M TECHNIQUE

FIGURE 2.4.8 R-H TECHNIQUE

These highly refined techniques of basic oxygen steelmaking are complex, and are described in detail, including sketches demonstrating rates of various reactions taking place, vacua used, gas flow rates and ferroalloy consumption, by Yamamoto[14] and Kohno.[15] Their paper "Combination of constant medium-carbon end-point BOF operation and massive RH treatment" includes a summary of their results over an output of 550 000 tonnes p.m. from three BOFs and two RH reactors.

2.5 CASTING METHODS

2.5.1 Ingot casting

On completion of the refining operation, the liquid steel with specified C, Si, Mn, S and P, but also residual oxygen in excess, is poured into large ladles for casting into ingots. Generally a sample is taken at this stage for confirmatory analysis and precise classification and some deoxidants are added. Casting ("teeming") is effected by running liquid steel at a controlled rate and temperature into moulds of appropriate shape; methods of teeming practised, are shown in Fig. 2.5.1. The moulds are usually made from cast iron and will have been cleaned and dressed internally before use so as to minimize surface faults. Ingot weights will vary according to the equipment available, and can be in excess of 20 tonnes.

In addition to the possibility of having surface faults due to splashing, etc., the steel does not solidify uniformly, but suffers from segregation of its carbon, phosphorus and sulphur parallel to and at right angles to the ingot length.

It will also suffer from physical defects such as blowholes and pipes, slag inclusions, etc. When molten steel cools to its solidification range the solubility of gases present decreases, causing a major part to be expelled, due to the change of the equilibrium state of the reaction:

$$[C]_{Fe} + [O]_{Fe} \rightarrow CO_{gas}{}^{*}$$

Much of the expelled gas will be entrapped into what are termed blowholes, as the steel solidifies. The form and distribution of the defects will vary according to the type of ingot, its shape and size, residual gas content on casting, and the conditions of teeming; a significant difference will also be found between the top and bottom half of the ingot (it is general practice to divide the ingot into two

*$[\ ]_{Fe}$ = in solution in iron

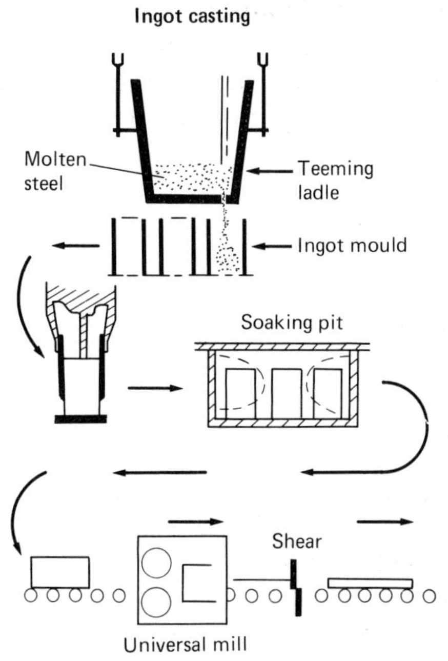

FIGURE 2.5.1 INGOT CASTING

halves, lengthwise, by cutting the slab produced in primary rolling into two short slabs). Ingot casting is a sensitive operation and demands considerable skill and experience.

Three basic types of ingot can be produced: killed, semi-killed and rimmed, dependent on the pretreatment of the liquid steel and the method of casting. The intermediate type semi-killed, is not used in the making of tinplate steels.

"Killed" steel is so described because there is no visible effervescence or gaseous reaction during solidification. This is because sufficient deoxidant, in the form or ferrosilicon, ferromanganese and aluminium (AK), has been added to convert the dissolved oxygen into insoluble manganese silicate and aluminium oxide. This combination is utilised because the silicate is liquid at the prevailing temperatures and agglomerates into large globules which rise quickly to the slag, carrying the very small alumina particles with them. This practice also limits residual silicon, which may adversely effect the corrosion resistance of the steel. As the killed steel solidifies, because solid steel has a higher density than liquid, it forms a contraction cavity — or "pipe" — at the top of the ingot; this portion will include a number of defects in addition to the void, and will need to be removed by sufficient cropping-off of the slab end after primary hot rolling. This will result in a lower yield, and the product is thus more costly (Fig. 2.5.2).

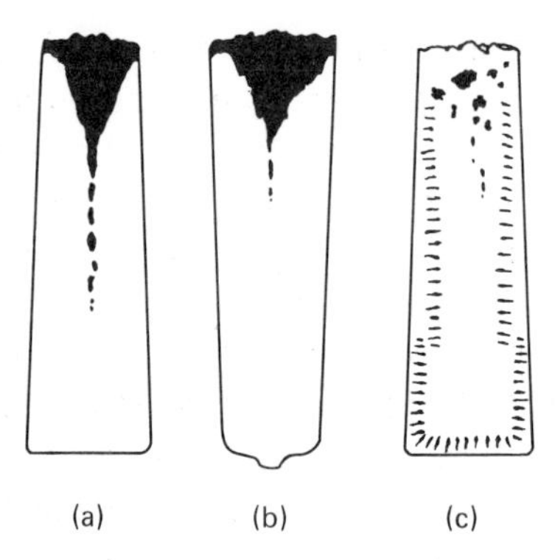

FIGURE 2.5.2 TYPES OF INGOT: (a), (b) SOLID OR KILLED, (c) RIMMING

Liquid rimming steel will contain a carefully controlled amount of oxygen after the addition of the same deoxidants in smaller amounts and when solidification starts, the solid forming inside the mould will be very low in impurities and carbon. This will raise the concentration in the liquid and a point will be reached at which evolution of CO will take place with some vigour. This is allowed to continue, to give the required rim thickness, and it is then stopped by placing a heavy metal cap on the top and freezing the surface. This causes the pressure to rise in the liquid and suppresses the CO-forming reaction. Solidification of the residual homogeneous liquid is completed, including the formation of gas pores containing CO. The solid ingot cross-section shows an exterior shell of high purity iron, with a small number of blowholes normal to the mould wall, a thin zone, immediately inside this, in which there is a high concentration of sulphur and phosphorus and a very large number of small blowholes and a central portion approximating in composition to the deoxidised liquid. The blowholes are filled with CO and will weld up during hot rolling, and their volume is such that there is no central "pipe" or cavity. A "bottle-top" mould can be used as an alternative to capping, so reducing the amount to be removed by cropping after rolling to slab.

The rim persists after rolling to the thinnest commercial sheet and provides a ductile surface which is corrosion resistant and readily coated with tin in an adherent and continuous film. A rimmed/capped (RC) ingot is likely to show less foreign matter than a fully rimmed type and also less segregation. It is also considered that the bottom half of the ingot will in general be somewhat cleaner that the top half, and although this is not always the case, it is the practice sometimes to use the bottom half only for more critical applications — the slab rolled out from an ingot of up to 25 tonnes is generally cut into two for further processing. This type of ingot is favoured, as its pure rim provides a clean defect-free surface to the ultimate sheet; it is relatively less expensive and overall a fairly clean steel.

With increasing use of the DWI and DRD processes, the demand for very clean steel is increasing appreciably, and will call for wider use of the latest techniques for close expert control; the need for continuous casting will also increase.

With the increasing attention to clean steel, methods of accurately assessing the overall incidence of NMI in a batch of steel have been developed; but all require detailed and somewhat painstaking application. General reviews of the occurrence of NMI in low carbon steel have been widely published; the following are particularly useful:

> 75th BISRA Steelmaking Conference[9], and BSC/BISPA Conference 1973[10].

Two basic types of assessment methods are used, metallographic and magnetic; both include a means for reporting the findings quantitatively. Moore[16] and others have developed a method based on a count of the NMIs of given length per 100 m^2 as a function of NMI length in the rolling direction; they found that there seems to be a threshold to distribution of NMI sizes, about 25 μm for AK steel and 80 μm for RC steel, referring mainly to so-called micro-NMIs. The limitation in quantifying macro-NMIs seems to have been largely overcome by the development of a radiographic technique (J.H. Wilson, BSC Tinplate Research, unpublished) and by the use of a magnetic particle technique (J.H.R. Lodge, BSC Tinplate Research, not yet published). Both of these are complementary, as their resolution is confined to NMIs of greater than 200 μm. The magnetic particle method has been extensively developed and is non-destructive; it has been licensed for manufacture as the Magmid by the Boyle Industrial Gauging System Ltd.

2.5.2 Continuous casting

The "casting steel into ingots" route requires "soaking" to a comparatively high temperature, suitable for hot rolling into slabs in a primary mill, before hot rolling into strip can be carried out, a major and costly operation. Continuous casting (CC) of the steel directly into slabs eliminates both of these operations; some metals have been continuously cast for many years, but the method has been developed to large-scale operation relatively recently for low carbon steel. The problems have been associated with its high melting point, high specific heat, and low thermal conductivity in comparison with other metals.

Continuously cast tinplate-type steels have been available for some years. Among the many advantages (Fig. 2.5.3) of this technique, the following are the most important:

(a) appreciable energy savings;
(b) considerable improvement in slab yield, with lower scrap;
(c) elimination of ingot casting, reheating and primary rolling provide significant savings in operating and capital costs;
(d) slab is more homogeneous and contains lower amounts of NMI than ingot steel;
(e) faster throughput.

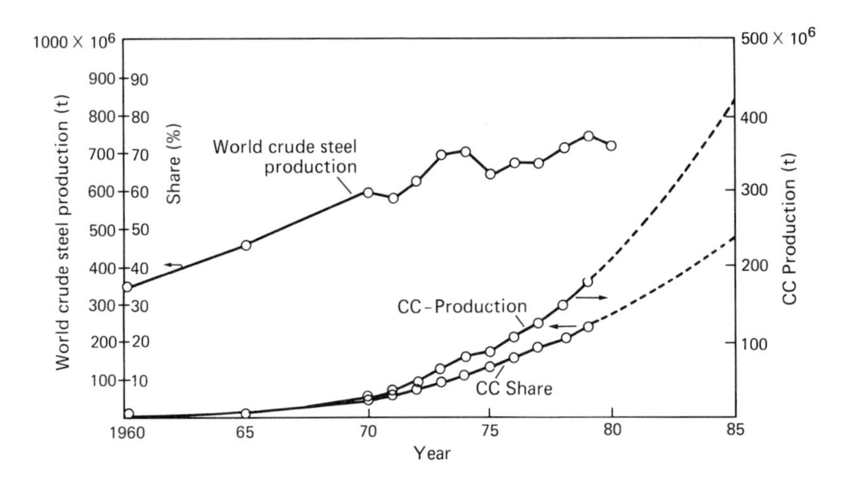

FIGURE 2.5.3 DEVELOPMENT OF WORLD RAW STEEL PRODUCTION AND CONTINUOUS CASTING OUTPUT SINCE 1960

Definition of slab yield can vary, and often the claimed improvements need to be treated with caution, but the following can be regarded as generally attainable:

Slab Yield %

Type	Ingot	Continuously cast	Approx. saving
Stabilised	84	92+	9%
Rimmed/Capped	89-91	–	3%

In addition to the definition used, actual yields can vary significantly between plants.

Metallographic examination has demonstrated that the number of NMI is realistically reduced, although some authorities will state that the improvement is in the form of a more consistent cleanness throughout the batch, i.e. is less variable, the inclusions are finer and more widespread; the large inclusions seen in stablised ingot steel are said to be almost unknown in continuously cast slabs. The greater uniformity is clearly demonstrated by a narrower spread in HR3OT hardness values, both longitudinally and across the slab. These advantages are even greater when combined with one of the secondary steelmaking processes described earlier; but this is a complex process and considerable skill and close attention to detail are required to achieve the high standards that are possible and indeed required by modern canmaking techniques, if efficient production and quality assurance are to be regularly obtained.

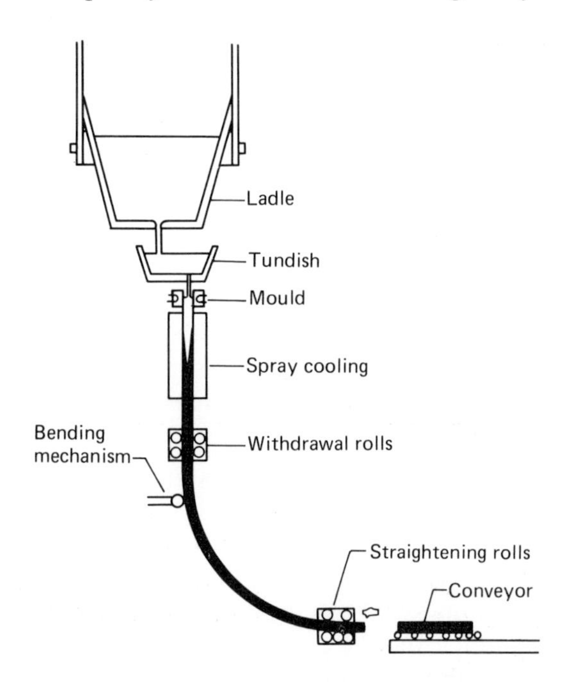

FIGURE 2.5.4 BENT STRAND CONTINUOUS CASTING MACHINE

It has been estimated that the production of low carbon steel by continuous casting — separate figures for tinplate steel do not appear to be available — had risen to about 30% overall by 1980, but the level will vary widely between countries and organisations. Japan had by that time reached a production of at least 60%, and some European countries had also increased their output substantially; the decision to convert to CC is complex, depending on such factors as local conditions, type of existing plant and the standard already being achieved. Estimates of ultimate production level vary widely, but it seems reasonable that worldwide an output of 60% of the total will be obtained, with some areas reaching nearly 100%; in fact some integrated plants use only continuous casting.

The principle of the process is shown in Fig. 2.5.4. The ladle, which may be supported at a height of more than 30 m above the main floor, will contain several hundred tonnes of molten steel at a temperature in excess of 1500°C. This is run through a refractory transfer tube at a carefully controlled rate into the tundish, the capacity of which will be in excess of 25 tonnes. From here it passes through nozzles into the mould, which is made of copper and is water-cooled (Fig. 2.5.5); the size will depend on the slab required, but typically it will have a section of about 1250 × 250 mm and a height of 900 mm. A starting block is fitted into the base, to close the open mould when casting starts. As the steel passes through the mould, a thin solid skin will be formed before it emerges. The mould is oscillated vertically through a small amplitude as any sticking to the metal surface will tear the weak skin. Water sprays under high pressure will complete solidification of the steel; the steel will have been "killed" with ferroalloys and aluminium in the ladle, as it is essential to avoid any gas evolution when only a weak skin has formed. In addition to the usual non-metallic inclusions, the steel can also be contaminated with particles of refractory material from the tundish; this calls for regular treatment of the interior between casts. Great care is also necessary to avoid oxygen pick-up, and this is ensured by immersing the quartz nozzles below the liquid steel in the mould and by covering its surface with a layer of suitable inert powder.

Speed of casting will vary usually between 1-1½ m/min. Originally, providing the number of slab widths required for hot-rolling programmes was a major problem, but this has been overcome by the development of a mould design which can be adjusted automatically for width in fairly small intervals.

As the lower half of Fig. 2.5.4 illustrates, many casters are capable of bending the vertically cast slabs through 90° into a horizontal position; this is achieved by means of a number of pairs of rollers as soon as the steel has solidified. Computerised control of the whole complex operation has, understandably, been adopted in many areas in Japan and Europe.

When the steel has been cast into ingots, these are stripped from the moulds and after reheating to about 1250-1300°C are rolled into slabs similar in dimensions to the continuously cast slabs, up to

250 mm (10 in) thick and 1000 mm (40 in) wide. Ingot rolling is carried out in heavy duty slabbing or universal mills, incorporating sturdy vertical edging rolls to control the width of the slab; these are sheared into appropriate lengths, usually two from each ingot. Slab ends are also sheared to eliminate all casting defects (pipes and significant amounts of non-metallic inclusions).

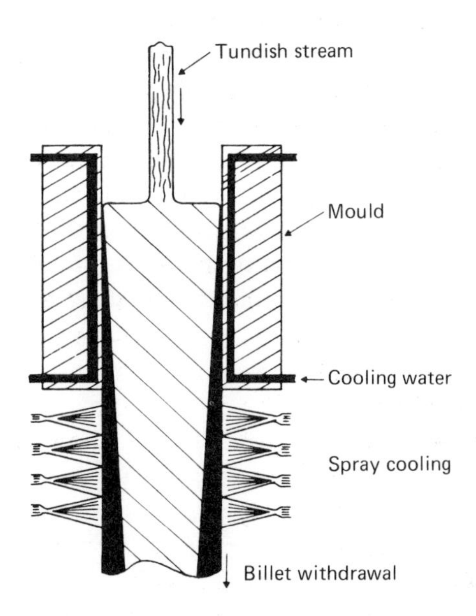

FIGURE 2.5.5 CONTINUOUS CASTING MOULD

Inevitably some defects will be present on the surface of both types of slab; these must be removed to ensure that ultimate surface quality will be adequate. After careful inspection, the areas containing faults will be "scarfed" by hand locally, using oxy-gas burners, or the slabs may be fed through automatic scarfing machines, in which all surfaces will be flame-treated. The repair is sometimes effected by mechanical gouging or chiselling.

2.6 HOT ROLLED STRIP PRODUCTION

Reduction of the steel slab to the final sheet thickness is carried out in two stages:

1. Hot rolling in a large continuous hot mill to a thickness of about 2 mm (0.08 in); this produces a very long strip of steel, which has to be coiled for interstage handling. This hot rolling operation must ensure that a good strip shape, internal struc-

ture and surface quality are achieved, as they affect the quality which it is possible to obtain in almost all following stages. All rolling conditions must be closely controlled — precise and uniform temperature control is also vital for many reasons as discussed below. Substantial weight loss and considerable surface roughening can occur because of oxidation at the high rolling temperature.

2. Cold rolling to the final thickness specified, also in a highly sophisticated mill. The advantages of cold rolling as the final operation are set out in Section 2.8.

Precise temperature control is vital throughout the hot mill; this is ensured by temperature control on entry of the steel, immediately after rolling, and the rate of cooling along the long run-out table to the coiler; this will ensure a strip of good shape and uniform thickness, as well as the desired metallographic structure.

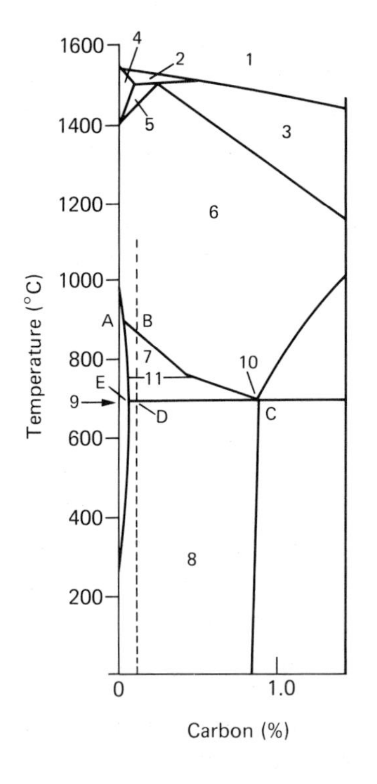

1. Liquid;
2. Solid Solution, in which Iron is in the ς Form, + Liquid;
3. Solid Solution, in which Iron is in the γ Form (Austenite), + Liquid;
4. Solid Solution in which Iron is in the ς Form;
5. Solid Solution in which Iron is in the ς Form, + Solid Solution in which Iron is in the γ Form (Austenite);
6. Solid Solution in which Iron is in the γ Form;
7. Solid Solution in which the Iron is in the γ Form (Austenite), + α Iron (Ferrite);
8. α Iron (Ferrite) + Pearlite;
9. α Iron (Ferrite) + Pearlite;
10. Eutectoid (Pearlite);
11. Magnetic Change Point (Curie Point).

FIGURE 2.6.1 IRON-CARBON EQUILIBRIUM DIAGRAM

The metallographic structure, and therefore the mechanical properties of low carbon steel, can be markedly altered by changes in temperature; its equilibrium diagram (Fig. 2.6.1) shows the limits of composition and temperature within which the various phases or constituents of the steel are stable. Changes in structure and the proportions of the constituents can be determined from the diagram. Our area of interest is that up to the dotted line at about 0.1% carbon. At room temperature and under equilibrium conditions this steel will consist of two phases: α-iron, termed ferrite (virtually soft ductile pure iron), having a body-centred cubic structure, and pearlite, a physical mixture consisting of fine alternate lamellae of cementite (iron carbide, Fe_3C) and ferrite (this is a relatively hard and brittle constituent). But, as will be seen later, the occurrence of pearlite in tinplate steel sheet is rare. Above about 870°C — the precise temperature depending on the total effect of all the elements present — and up to about 1440°C, the iron is in the face-centred cubic or γ form, capable of dissolving up to 1.7% C to give the single-phase solid solution known as "austenite", soft and ductile. Below 720°C, the iron is in the body-centred α form, capable of dissolving a maximum of 0.018% C at this temperature, and much less on cooling. Between 870° and 720°C is the "critical range" of temperature, within which α and γ can coexist, the proportion of the latter decreasing as the temperature falls. Below 720°C, any carbon in excess of the solubility at any temperature is present as iron carbide, Fe_3C, and in the most suitable structure it will exist as a uniformly distributed precipitate of small particles. The actual structure will depend on the rate of cooling through the critical range. If it is slow, there will be few nuclei of Fe_3C; these will grow to give "massive cementite", but if the cooling rate is increased, the desired structure can be achieved. This requires that the hot rolling is completed above the critical range, and that cooling to about 700°C is rapid before coiling. Under these conditions, the ferrite grains are equiaxed and uniform in size. The carbide distribution and size are as illustrated in Fig. 2.6.2. When stabilised steel is used, lower coiling temperatures, e.g. below 600°C, are required to hold the aluminium nitride in solution; this is particularly helpful to develop high *r* values (Chapter 5). Thus close control of temperatures is required during hot rolling, particularly to finish "high" and coil at "low" temperatures. Considerable effort is being applied to develop satisfactory cooling systems on the run-out table for precise control of coiling temperature.

The principles of rolling metal are illustrated in Fig. 2.6.3. The two rolls rotate with the same angular velocity, but in opposite directions rotationally. The gap between them controls the thickness of the rolled product. Pressure applied vertically causes the rolls to grip the metal and force it through the gap, thereby reducing its thickness and increasing its length correspondingly. Lateral spread (at right angles to the direction of rolling) is normally minimal at these thicknesses.

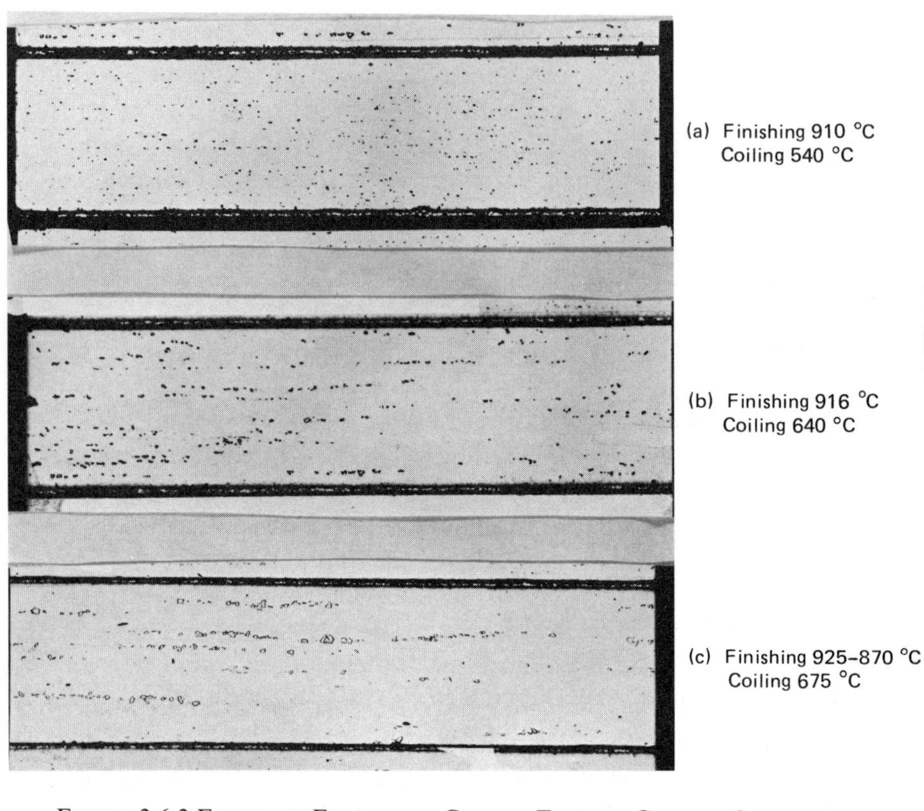

FIGURE 2.6.2 EFFECT OF FINISHING & COILING TEMP ON CARBIDE STRUCTURES

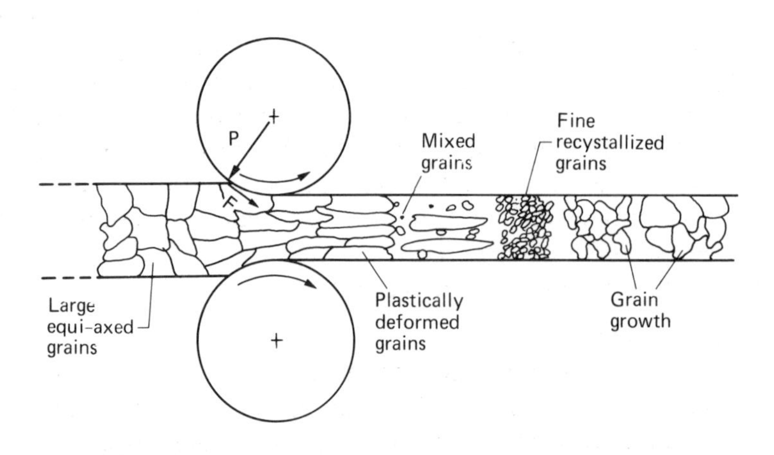

FIGURE 2.6.3 GRAIN STRUCTURE DURING HOT-ROLLING (SCHEMATIC)

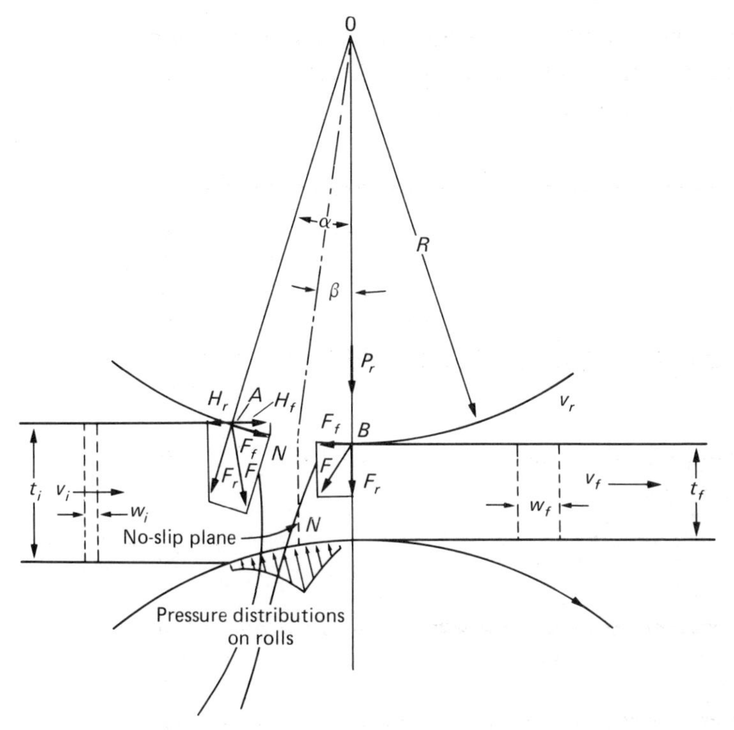

FIGURE 2.6.4 FORCES AND VELOCITIES DURING ROLLING

It is clear that the strip speed must increase from that at which it enters the nip of the rolls to the point where it emerges from them, and also that strip speed can only be equal to that of the roll surface at one point, the neutral point – (Fig. 2.6.4).

Two forces are operating, the radial pressure (*P*) and the tangential frictional force (*F*). Up to the neutral point, *F* is acting to draw the strip into the roll gap; but after passing the neutral point and as the strip speed is increasing, *F* will act in the opposite direction.

Roll pressure is not uniform, it is at the maximum at the neutral line; in fact the neutral is not along a line, but rather a narrow band. The shaded area in Fig. 2.6.4 represents the force required to overcome friction, and the area under the line *AB* represents the force needed to deform the steel.

The main parameters of rolling are:

the roll diameter,
the deformation resistance of the steel,
magnitude of the friction between roll surface and strip,
the front and back tension in the plane of the strip,
peripheral roll speed.

Analysis of the forces and the geometry can be summarised, in terms of their effects on the rolling load, as it:

(i) is decreased with decreasing roll diameter (is proportional to its square root),
(ii) increases with decreasing stock thickness, and
(iii) a point will be reached at which the resistance to deformation is greater than the roll pressure which can be applied; this is limited by the severe elastic deformation of the rolls under the heavy load.

Thus smaller diameter rolls allow a greater reduction in thickness before roll flattening becomes excessive — hence the optimum system will be small diameter working rolls which will be adequately supported against deflection by heavy back-up rolls.

It also demonstrates that the higher the friction, the greater will the rolling load need to be.

The technology of hot rolling and mill design have been improved substantially over the last 20 years or so. A review by Keefe[(17)] of the improvements made up to 1979 and the range in mill designs in use provides a comprehensive picture of these developments; it covers methods of improved shape control, gauging equipment, thickness control and better roll wear, higher mill loading and greater speeds and several other aspects. It concludes by pointing to the further research and development needed to improve material quality and mill operation. The paper was part of the proceedings of a conference arranged by the Metals Society on Flat Rolling held in 1979. Its emphasis on the need for continued extensive effort to continue improvement further was highlighted by a most comprehensive international conference on steel rolling organised by the Iron & Steel Institute of Japan in 1980.[(18)] It was presented in 110 contributions by specialists from many countries, the proceedings covering some 1000 papers. The two broad fields, progress of rolling technology and contributions to rolling research, covered practically every aspect of the technology, production techniques, shape and thickness control, advanced computer control of setting-up and operation and current research.

The need for good shape and thickness control to allow the next stage, cold rolling, to be carried out as precisely as possible is emphasised in many papers; increasing rolling speeds add to the complexity of these controls, and to the need for the operator's expertise to be complemented by precise and rapid physical aids.

A modern high-speed mill will contain, in addition to the mill housing and roll assembly, a number of vital sophisticated control units; very large electric motors supply the considerable power required, and these will be operated by control units to govern speed and between-stand tensions, to enable the mill to run at variable speeds; screw-down mechanisms or hydraulic units applying the roll loads will have associated load cells for precise measurement of applied loads; pyrometers to measure temperature at all relevant points along the whole mill; radiation or x-ray thickness meters; and accurate speed indicators. Modern mills are complex large installations, rapid and precise, and computer control is vital for set-up and operation. Modern control units will include shape and profile control and means for controlling roll gap contour.

Modern mills will normally include a two-high scale breaker, followed by a roughing section; consisting of 4-6 four-high roughing stands, and a finishing section, generally with 5 or 6 four-high stands. Copious water sprays are employed to remove scale and to provide adequate means for controlling strip temperature, the latter extending along the long run-out table to the coiler (Fig. 2.6.5).

FIGURE 2.6.5

2.7 OXIDE FORMATION AND REMOVAL

Substantial layers of oxides are formed when the steel is heated to elevated temperatures for rolling, and these need to be removed completely before proceeding with the next stage, cold rolling, to avoid the defects that would otherwise occur to the strip surface. The structure of the oxide layers formed under these conditions is usually taken to consist broadly of three layers:

the outer thinner layer	Fe_2O_3	(haematite) 70% Fe
the middle thicker layer	Fe_3O_4/FeO	(magnetite/wüstite) 72% Fe
the innermost layer, the thickest	"FeO"	(wüstite) 75-77% Fe

The precise composition of the "mill" scale will vary appreciably according to the temperature and times involved, the rolling practice and the composition and physical characteristics of the steel; the oxides do not form discrete layers, but diffuse into each other. Fe_2O_3 is relatively insoluble in mineral acids, and is generally considered to be removed by disruption due to the acid attack taking place under the film. The total thickness of the scale will also vary, but will approximate to 0.01 mm.

The process widely used for the removal of scale is termed "pickling", which derives from the old practice of immersing the sheets for a long time in very dilute acid solution. The modern process is not costly and is adequate for cleaning the surface efficiently, preparatory to further processing. The long established pickling liquor is a dilute aqueous solution of H_2SO_4 (10-15%) used near boiling point. Resulting products are ferrous sulphate and gaseous hydrogen. In more recent years hydrochloric acid, used at a much lower strength of 5-10% HC1, and at temperatures slightly above ambient, has become very popular. It is able to dissolve haematite and magnetite also to a degree, thus is more versatile, is faster in operation and is less inhibited by build-up of metal salts. It is thought to be more effective in scale removal and leaves a cleaner, brighter sheet. Appropriate inhibitors are added to limit attack of the steel, giving a smoother surface and some economic advantage. In each case, thorough rinsing and drying are essential to achieve acceptable surface quality.

Scale removal is assisted by breaking it off by means of a roller leveller installed in advance of the pickling bath, in the continuous pickler. The strip is usually passed through hot water to preheat it prior to entering the acid bath. Overall, there is a significant loss of steel by scaling, often over 5% over the two hot rolling stages. Removal of the surface layers by scaling and "pickling" is sometimes regarded as a beneficial loss, as some of the surface blemishes are thought to be removed.

Passing strip through the automated pickling line gives an opportunity for inspecting its quality, thickness and width at entry, as well as the surface finish after leaving the drier. Some defects, due to rolling or casting faults, are sometimes more readily seen after pickling. Overpickling, which can readily occur during line stoppages, can cause blisters, pits or larger rough areas, which will prevent effective coverage of the steel and disruptions in the metallic coatings subsequently applied. Underpickling will leave areas of scale on the surface, resulting in similar incomplete coating. For similar reasons, good rinsing and drying are vital to provide adequate surface preparation; means are also provided to minimise drag-out from the pickling baths.

Detailed discussion of pickling processes are given by Hoare(19) and Gabe(20).

At the exit end, the strip is side trimmed to the exact width required. This trimming at the same time removes any serrations, cracks, or other bad edges which may have occurred during hot rolling. A film of oil is then applied to the surface to assist coiling, to help preserve the pickled coil during storage, and to act as a lubricant during rolling in the first stand of the cold mill.

2.8 COLD REDUCTION

The final stage of thickness reduction is carried out by cold rolling, as it confers several advantages over hot rolling for thin strip and sheet:

(i) Surface quality is greatly superior in respect of smoothness and appearance.
(ii) Better dimensional tolerances are achieved.
(iii) Cold rolling can be employed to increase plate strength (a typical example is double reduced plate, for which a different process sequence is used, and is described later).

Strip from the hot mill is normally within a thickness range 2.0 to 2.3 mm and as the bulk of tinplate gauges are within the range 0.16 to 0.28 mm the percentage reduction in thickness is very high, generally between 85 and 92%. The operation is usually referred to as cold reduction in the trade, and very robust mills having a large power supply must be used. As with the hot rolling mill, they are universally of the four-high type; the inner pair of working rolls have a diameter in the range 450-585 mm (18-23 in), and large, heavy back-up rolls up to 1300 mm (50 in) diameter. This combination of roll design, coupled with the provision of high strip tension, provides near optimum rolling conditions and near minimum rolling load down to the minimum plate thickness required.

The tandem mills used for cold reduction usually contain 5 or 6 four-high mills (illustrated in Fig. 2.8.1). The speed of the strip emerging from the last stand can be up to 2000 m (6500 ft) per minute. Latest mill designs have a very high output capacity, and require massive capital expenditure; operating costs, particularly that of the labour needed, are attractive, but they suffer from lack of versatility.

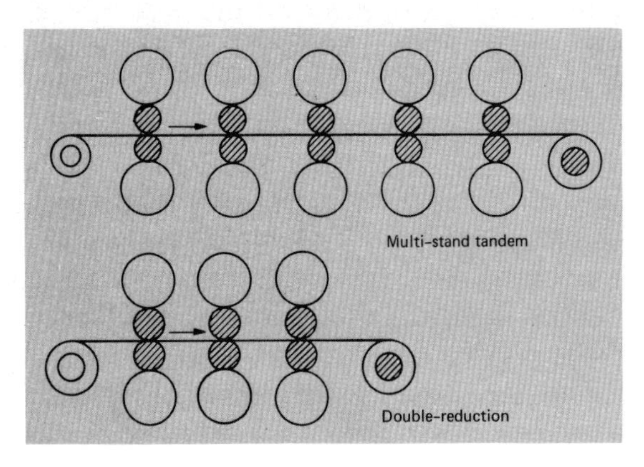

FIGURE 2.8.1 MULTI-STAND TANDEM MILLS-ROLL ARRANGEMENTS (DRIVEN COMPONENTS HATCHED)

Four-high reversing mills have good flexibility and are also capable of being automatically controlled; their operating speeds, however, are rather low, and they have been used only in one or two areas for tinplate base production. The design of a modern four-high reversing mill has been described by Booth[21].

Large amounts of heat are generated by the heavy cold work involved, the emerging strip temperature can approach 200°C, and a very efficient lubricant/cooling system is vital to maintain friction at a low level (μ=0.05 – 0.10) and to extract the heat from the rolls and strip. Various lubricants are employed, usually dispersed in water, or the water may be applied separately on to the rolls. Often, different mixtures are used on the first, middle and last mills in the train. Efficient operation of the mill is greatly dependent on adequate cooling and lubrication; in addition to productivity and general strip quality, roll wear and strip shape in particular can be badly affected. Circulatory systems with filtering devices are usually employed, and the lubricant properties are checked regularly. This important aspect is discussed in detail by Hoare[19] and Earnshaw[22].

In operation, the strip is paid off the coil loaded on to the uncoiler, and fed slowly through each stand and finally on to the recoiler mandrel (the wind-up reel); the mill is then run up to maximum speed, and finally decelerated as the tail end of the coil is reached. A new coil is then fed on to the uncoiler and the processes repeated; thus cold reduction is a batch process, and substantial operating time is lost in between coils due to the lower speed during run-up and run-down, and a significant length at the front and back ends of the coil will be outside gauge thickness tolerance.

The tandem mills are equipped with a number of instruments to measure roll and strip tension forces, strip thickness after the last stand (usually by X-ray gauge) and various mill data such as strip and roll speeds, coolant flow rates, roll pressure, screw-down position, etc.

Two of the most important plate characteristics for high-speed canmaking processes are close gauge tolerance and good shape ("flatness"); deviations in both can originate in the hot mill and the best cold rolling practice cannot hope to remove them completely. However, a great deal can be done in the cold mill to produce strip which is good in both respects; full control of all the numerous mill features is difficult by the operators alone, and many physical aids have been developed to assist them, similar to those in hot rolling, to produce the even higher quality standards being demanded by modern processes.

A joint development by Hitachi Ltd. and Nippon Steel Corporation in Japan has produced a six-high mill — the HC mill — in which a set of rolls is fitted between the back-up and the work rolls; it is claimed that this system materially assists production of flat strip and reduces the drop in strip edge thickness to about half. A further advantage is that it eliminates the need to crown the rolls to counter uneven sheet contour; [23,24] Fig. 2.8.2 illustrates its principle. It has been applied to upwards of thirty continuous mills, with beneficial results.

As the strip appears flat to the mill operator — being under high tension it is elastically stretched — a physical method has been developed by Davy-Loewy[25] of detecting uneven shape by on-line measurement of uneven tension locally across the strip. This has proved to be very useful on fast aluminium rolling mills, and is being applied to high-speed steel tandem mills. An on-line non-contacting shape meter has been developed jointly by the Ishikawajima-Harima Company and Nippon Steel[26] primarily for use on

Plant	Type of application	Purpose and advantage	Work Roll diameter mm
Kimitsu Works, Nippon Steel Corp.	No. 6 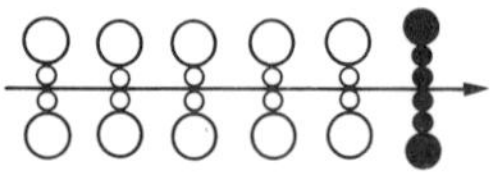	Shape control. Improvement of quality and yield. Production of thinner and harder strip.	4H-MILL 585 HC-MILL 535
Yawata Works, Nippon Steel Corp.	Nos. 1 2 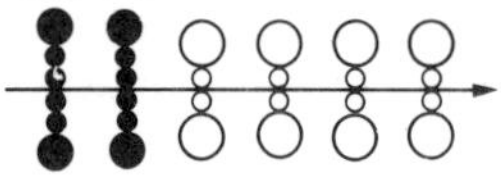	Thicker material can be used by heavy reduction. Energy saving at hot rolling and pickling line. Productivity boost Excellent shape and reduction of edge drop.	4H-MILL 588 HC-MILL 435
CSN (Brazil)	No. 1 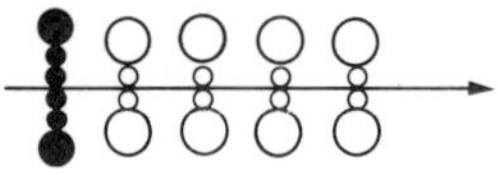	Thicker material can be used by heavy reduction. Energy saving at hot rolling and pickling line. Productivity boost Excellent shape and reduction of edge drop.	4H-MILL 483 HC-MILL 420
Nagoya Works, Nippon Steel Corp.	No. 1 No. 6 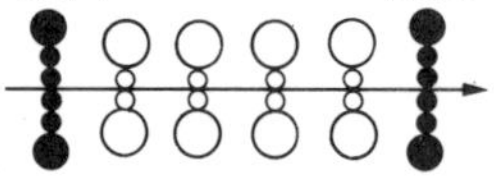	As desribed in items (1) and (2) above.	4H-MILL 585 HC-MILL 440

FIGURE 2.8.2 SIX-HIGH MILLS

hot mills, as this is regarded as being more important with respect to final shape than the cold mill operation. It is based on an optical method, a camera recording continuously the image of a "bar-type" incandescent lamp on the steel strip surface during rolling. Experience has shown this also to be an effective means of improving "shape" quality.

Many devices for reducing variability in gauge have been in use for some years, with varying success; there is little doubt that gauge uniformity has been considerably improved in recent years by the use of automatic gauge control (AGC) units. The Davy-Loewy hydraulic AGC system, for which considerable success seems to have been achieved, is described in detail by Dendle[(27)] with examples of its effectiveness in reducing gauge variability. Further data on its performance are given by Hobson [(28)].

Extensive development work has also been carried out for some time to improve overall mill operation and quality of strip, by automatic control of all relevant mill factors. Many well-proven systems are now in use; these appear to have been assisted in improving productivity and product uniformity by operating with computerised control. They are used for rapid initial setting up of the mill and during continuous operation. Much literature has been

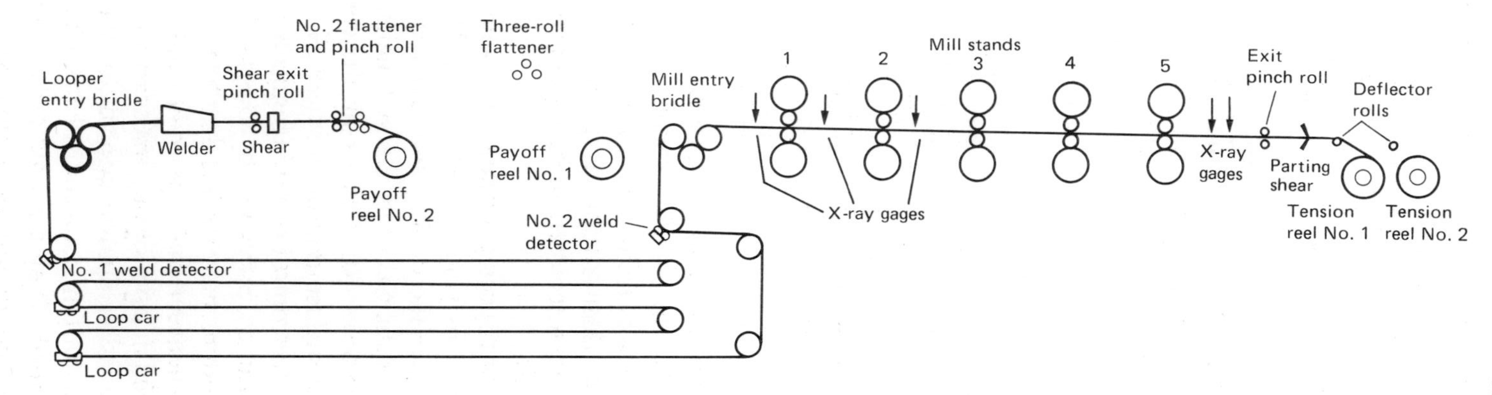

FIGURE 2.8.3 5-STAND CONTINUOUS COLD-ROLLING MILL LAYOUT

published on these devices, of which the following is a selection: Bryant,[29] Kalman,[30] King,[31] Kopp[32] and Ter Maat.[33,34] The latter gives a detailed account of the Estel Hoogovens systems applied to their "tinplate" cold mill and their experience of its advantages over a 2-year period. In particular, it is shown that thickness tolerance has been substantially improved (to less than 1%) and that this has been accompanied by a reduction in total rejectable off-gauge strip length from a typical 120 m to 20-40m/per coil. Most of these improved control systems can be applied to established mills.

As mentioned earlier, the multi-stand tandem mill is in fact a batch process, as one coil on the mill is completely passed through and recoiled before the next can be threaded through slowly by hand and then rolled at speed. Thus, appreciable production time is lost between coils; Weirton Steel[35] have developed a truly continuous – system wherein the back end of the coil within the mill is trimmed and butt welded to the front end of the following coil whilst it is running at about 1200 m (4000 ft) per minute; joining is carried out within 90 sec, and the mill is then accelerated to its full speed, about 1980 m (6500 ft) per minute. This short welding time is made possible by incorporating a 300 m (1000 ft) capacity strip accumulator. These modifications were made to a tinplate gauge cold mill which was also automated and computerised as comprehensively as possible. Its description includes a schematic diagram of the mill; this is reproduced in Fig. 2.8.3. Experience to date indicates real improvement in productivity, yield, quality and operating costs.

A review of modern cold rolling mills was presented in 1978 by Morris,[36] covering (i) how methods have progressed to the state of the art at that time, (ii) the developments foreseen in the future for cold reduced steel; it includes mills for tinplate. A more recent review, as reflected in the work of the VDEL Cold Rolling Committee, has been given by Schmitz[37] covering shape meters, surface inspection, the Hitachi-NSC "HC" mill, the variable crown rolling system, and sheet surface improvement.

Modern computer control of rolling mills is very complex, and is generally regarded as providing a complete system. Automatic gauge control systems have reduced substantially thickness variability within and between coils. In addition to the first x-ray thickness gauge situated in front of the first stand heralding the approach of a weld and providing appropriate adjustment to roll pressure, it forms part of a constant gap control system which compensates virtually instantaneously for many drift changes, such as strip tension, back up roll eccentricity, lubrication flow, roll bite friction, roll temperature and in fact all conditions having a significant effect on the rolling process.

Modern mills are often controlled by three computer units, all being interlinked:

automatic mill set-up, for changes in plate specification, which will include coil width and outside diameter, coil handling operation, threading, strip tail-out, initiation of the entry, run, and automatic slowdown, and of the joining welding cycle;
mill operation — automatic control of all rolling factors (basically roll pressure and mill speed/ acceleration in relation to gauge);
delivery end — sensing of coil-end with strip tension adjustment and slowing of speed, automatically shearing of coil end, and coil removal on to delivery conveyor.

The scheduling-automation computer passes the message signals to the process-automation computer, which will then monitor the whole process, and will include display of all relevant information for information and logging at many points. These comprehensive systems are vital to increase efficiency and reduce down-time, reduce extent of off-gauge, and increase throughput.

2.9 ANNEALING

2.9.1 Introduction

The effects of the heavy cold working (which is defined as plastic deformation in a temperature range within which the strain hardening it causes is not relieved) of the steel during cold rolling (not less than 85% reduction in thickness) are substantially to increase its strength and hardness, but at the expense of ductility; these are illustrated in Fig. 2.9.1.

Its effects on metallic structure have been discussed in some detail by McLean.[40] The considerable energy expanded on rolling is largely dissipated as heat, but part — possibly about 10% — will be transferred to substantially increasing its internal energy; most of that will be in the form of a major increase in crystal dislocation density (this is discussed in Chapter 5). Studies have shown that the dislocation density level of 10^6–10^8/cm^2 in well-annealed low carbon steel will be increased on the cold reduction to some 10^{12}/cm^2. This cold worked crystallographic structure is not thermodynamically stable, and when its temperature is increased, its degree of instability will increase rapidly and will finally revert to its original relatively strain-free condition. The mechanical changes are illustrated in Fig. 2.9.2 and the metallographic changes in Fig. 2.9.3 (a) and (b), the former reflecting the elongated ferrite grains amounting to a somewhat fibrous structure resulting from the heavy cold reduction.

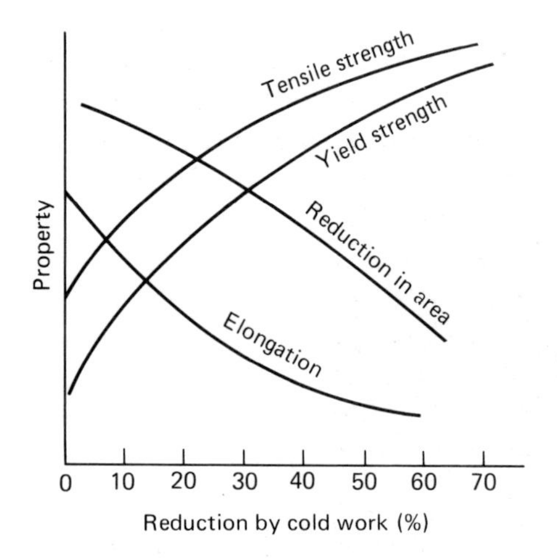

FIGURE 2.9.1 VARIATION OF TENSILE PROPERTIES WITH AMOUNT OF COLD-WORK

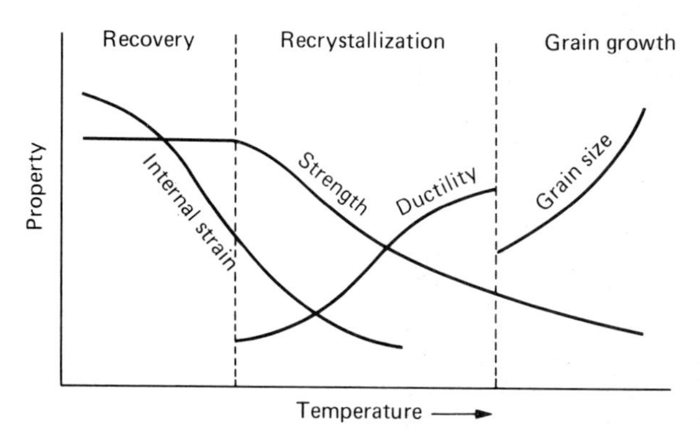

FIGURE 2.9.2 SCHEMATIC DRAWING INDICATING RECOVERY, RECRYSTALLIZATION, AND GRAIN GROWTH AND PROPERTY CHANGES IN EACH REGION

The first stage in Fig. 2.9.2 illustrates initial recovery of its physical properties, including almost complete restoration of its hardness, but without observable change in its metallographic structure; in the second stage, recrystallisation, the elongated grains are transformed into new fine grains, relatively equiaxed and strain-free. This is accompanied by marked increase in ductility and a corresponding decrease in strength. If the temperature rise continues into the third zone, an appreciable increase in grain size will occur resulting in the development of a coarse "orange-peel" surface on appreciable deep drawing, and should be avoided. Thus close control of temperature during this process annealing to eliminate the harmful effects of cold reduction is vital; the term subcritical indicates the temperatures are held below the critical range (Fig. 2.5.1). As these metallographic changes are time and temperature dependent, close control of treatment time is also necessary.

The structures of the steel as cold rolled and after batch annealing are shown in Fig. 2.9.3; the various ferrite grain sizes related to the annealing treatment are shown in Fig. 2.9.4.

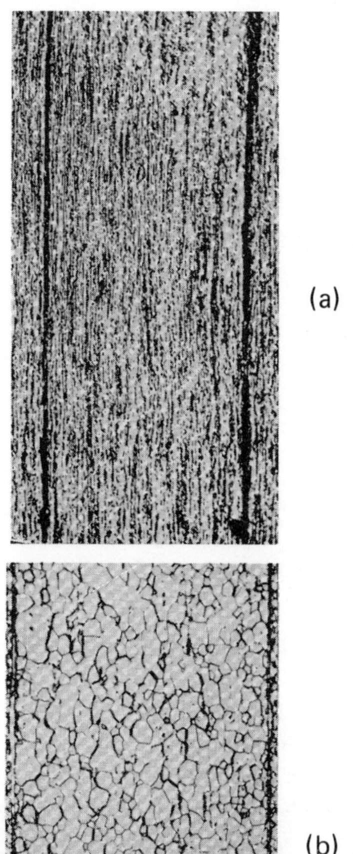

(a) As cold rolled

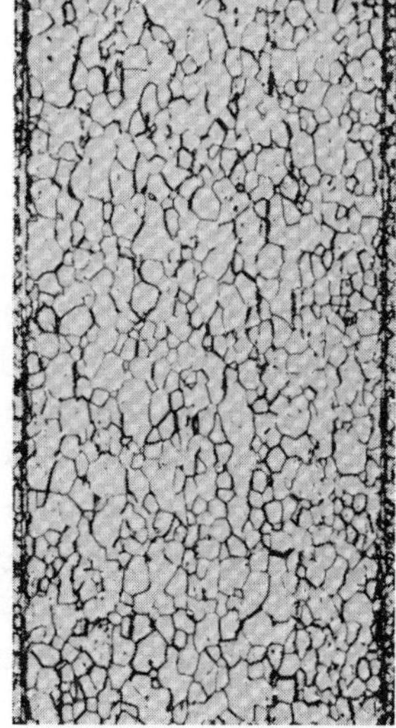

(b) After batch annealing

FIGURE 2.9.3 STRUCTURES OF THE STEEL (MAG x 130)

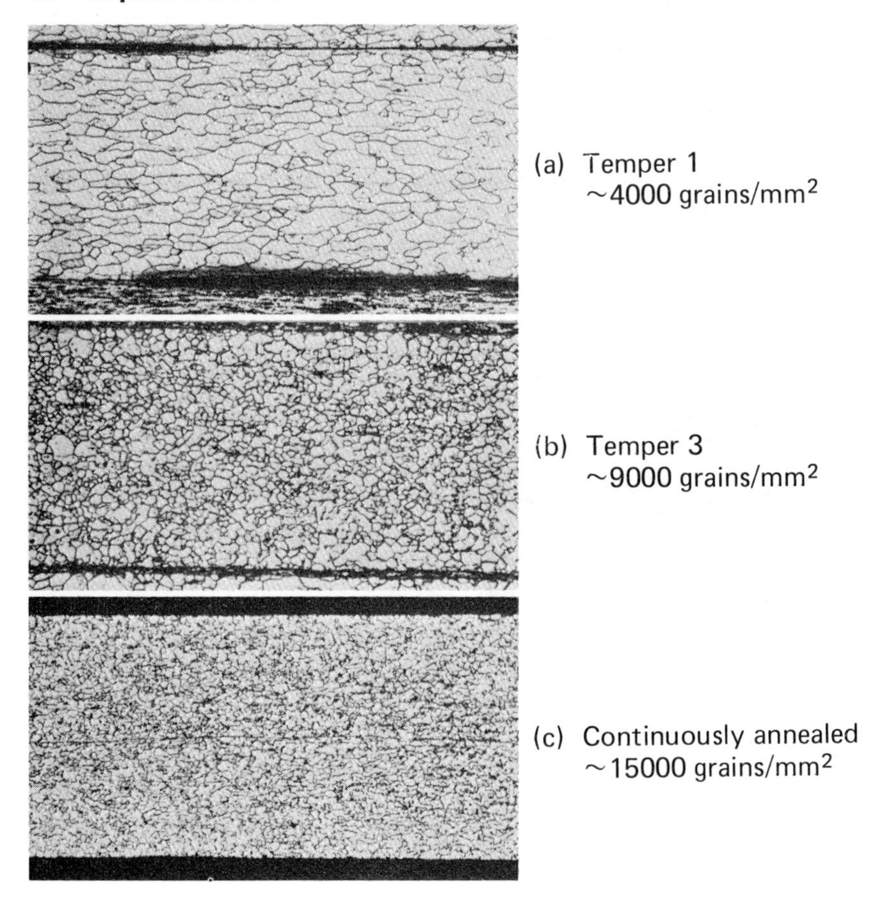

FIGURE 2.9.4 MICROSTRUCTURES OF DIFFERENT TEMPER GRADES (x 130)

2.9.2 Cleaning of strip surface

The lubricant flooded onto the steel strip during cold reduction will decompose at the high annealing temperatures used and will interfere with the effective coating with tin in the final stage; it is vital, therefore, that this is thoroughly removed before annealing. This is carried out in a separate continuous line when the batch (box) annealing method is used, but with the alternative continuous annealing it is usual to incorporate the cleaning treatment in an unit integral with the annealing line.

The usual method of cleaning is by electrolytic treatment in an alkaline solution. As the amounts of lubricant to be removed are large, strong solutions of caustic alkalis with phosphates or silicates and wetting agents are used as the detergent. Generally, cleaning is carried out in two stages, the first consisting of either a dip tank or the detergent is sprayed on to the strip, followed by thorough scrubbing and rinsing, then a similar electrolyte is used in which bipolar electrodes are fitted, again followed by thorough scrubbing and rinsing, and finally drying by means of warm air. Solution temperatures approaching boiling point are used, and close control of solution strength is exercised to ensure very effective cleaning. Modern lines are operated at strip speeds of the order of 1000m/min and the tanks are fitted with hoods to collect the fumes generated by the chemical actions.

U.S. Steel has developed a new high-current-density (HCD) technology for cleaning CR steel sheet and strip; three versions are available for various applications.

2.9.3 Batch or box annealing

This process was developed in the early 1930s for process annealing of long thin steel strip when wound into coil form. The furnace consists essentially of two parts, a fixed base and a cover, the latter being portable. The cover contains the burners for heating the charge, and is in effect the furnace; both parts have been developed to a high standard to ensure that the desired temperature cycle, consisting of three stages, heating, "holding" and cooling, can be achieved as precisely as possible.

The furnace cover is lined with refractory material, and has burners incorporated into it — either direct fired or indirect radiant — throughout the unit, as well as a number of thermocouples suitably positioned, here and in the fixed base; thus ensuring that the optimum metallographic structure and desired mechanical properties are achieved.

A base designed to support several stacks of coils is provided, and up to four coils can be stacked vertically, according to the height of the cover and the width of the coils; the base may carry one stack or up to twelve. Each of the supporting platforms in the base will be fitted with a fan for circulating the hot atmosphere to give as uniform a temperature as possible; temperature distribution is also assisted by fitting diffuser and convector plates at the base of the stack and between coils within each stack. Within each base, inlets are fitted to provide entry of a reducing atmosphere having a very low moisture content, which is circulated through the stack continuously. The most widely used reducing atmosphere, termed HNX, contains 95% N_2, 5% H_2 and has a dewpoint of −40°C. Each stack is covered by a light steel inner cover which is sealed into the base by means of sand or liquid; this will ensure that the steel surface is not oxidised during the cycle, and the gases will be circulated until the metal temperature is finally cooled to a "safe" point.

The total mass of steel loaded into the furnace can be very large, up to several hundred tonnes; the heating stage of the cycle must of necessity be very long, often up to 30 hours duration. Soaking times will vary according to furnace design, chemical composition of the steel, and the temper grade desired, between say 5 and 12 hours; it is also the practice to use slightly higher peak temperatures for the softer grades, say about 650°C, and the order of 620°C for the harder temper grades; these are approximate only and will vary according to many factors. By the same token, cooling times will be long, even though a form of accelerated cooling is employed. The total cycle times, therefore, are of the order of 4 days; to assist throughput, the furnace cover is removed and transferred to another base already loaded with coils awaiting annealing; the number of fixed bases installed will be greater than that of furnace covers for this purpose.

Many types of defect can occur during the annealing process, in addition to those due directly to lack of temperature control and improper reducing gas practice.

More recent developments are the highly sophisticated versions of the single-stack annealing furnace. Basically these also consist of a fixed base, a sealable inner cover, and the removable outer cover, with an additional inner hood which is fitted over the inner cover when the outer cover is removed. They are fitted with highly efficient circulatory systems, water-cooled heat exchangers and rapid cooling hoods.

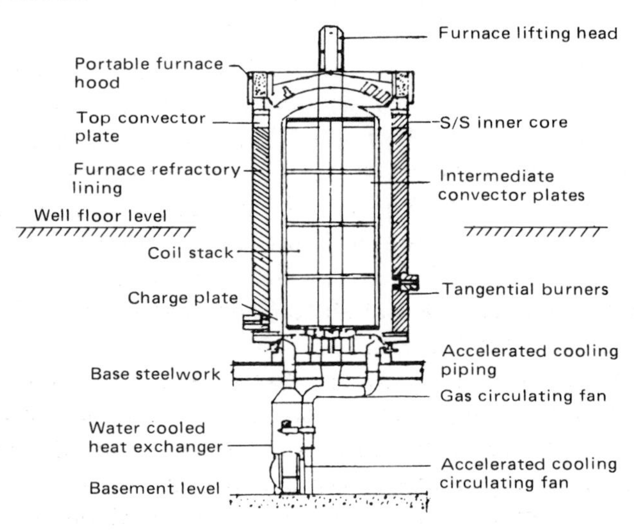

FIGURE 2.9.5 SINGLE-STACK ANNEALING UNIT

The entire heating/cooling cycle is computer controlled; heating rate is more rapid and the heating/cooling cycle is more precise and uniform than is possible in the batch annealing units. Although the maximum charge is much lower than in the latter, modern furnaces of this type can handle up to 100 tonnes. Fig. 2.9.5 illustrates a typical design.

2.9.4 Continuous annealing

As batch annealing is slow, causing a relatively serious hold-up in throughput, attempts were initiated by the Crown Cork & Steel Company in the United States in 1940 to speed up this essential process by installing a small continuous annealing line. This proved that rapid annealing was possible, and since that time it has been developed by many organisations into a fast, yet highly efficient, process. Basically it consists of passing a single strand of steel strip, hence the term strand annealing sometimes used, through a long furnace at speeds approaching 800m/min in modern furnaces. A section of a typical furnace is shown in Fig. 2.9.6.

As discussed later, the slightly different mechanical properties given by single-strand annealing, in comparison with those of the long-established batch annealing material, resulted initially in slow growth of the method. But as the process was improved further and canmakers gained experience in using the slightly more rigid material with a greater "spring-back" after forming, use of the method was accelerated substantially in the 1950s.

Total cycle time in the continuous annealing process is less than 90 seconds, in comparison with up to 4 days in batch annealing; a typical temperature/time cycle is shown in Fig. 2.9.7. It incorporates four zones, heating, soaking, slow cooling, and fast cooling. The peak temperature is usually 650°-690°C, but will be varied according to the earlier history of the steel, its composition and the temper grade required; e.g. Al killed steels for deep drawing require a temperature about 30°C higher than rimmed steel. The thin gauges obtaining in tinplate practice allow the adoption of short soaking times, of the order of 20 seconds, and yet provide sufficiently complete crystallisation.

Slow cooling over 30 seconds, down to a little under 500°C ensures uniform cooling free from variable stresses and distortion in the strip, and avoids any danger of quench aging. Beyond that temperature, cooling can be as rapid as possible.

The cleaning unit is situated in this case in front of the furnace and is integral with it. The strip is fed from the coil through the units

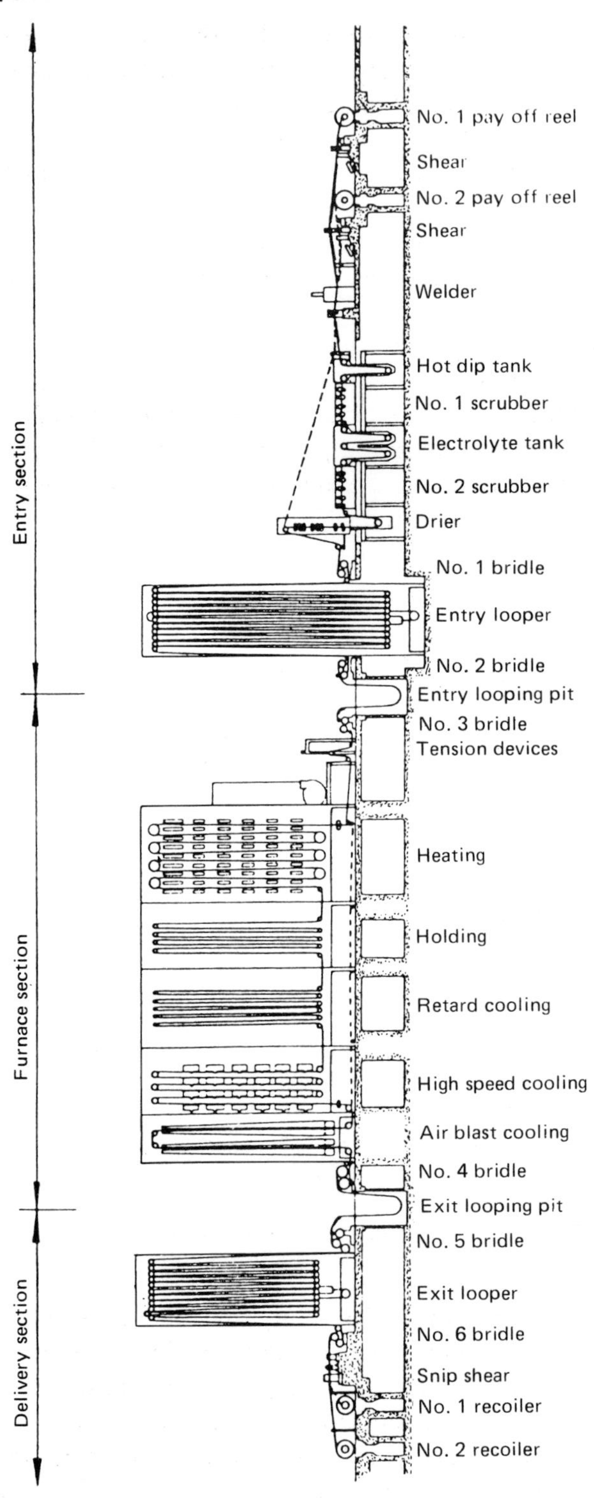

FIGURE 2.9.6 A TYPICAL CONTINUOUS ANNEALING FURNACE

horizontally. The first tank is normally a dip tank, followed by electrolytic cleaning or hot detergent sprays, effective rinsing, brushing and drying stages; it is, of course, basically similar to the separate line employed prior to batch annealing.

The strip then passes into the furnace, through which it is fed vertically, up and down, in a serpentine manner. It is vital that the strip does not stop within the furnace and, as with all continuous lines, a large reservoir is included to give ample time for welding the end of one coil to the leading edge of the next coil being fed through. Because of the high speeds of large modern furnaces, the length of strip in the looper alone can be up to 400 m. The entry and exit ends of the furnace have ingenious roll seal units which allow the reducing gas to be continuously pumped through without undue loss. Its composition can be mixtures of carbon monoxide and hydrogen; again frequently of HNX.

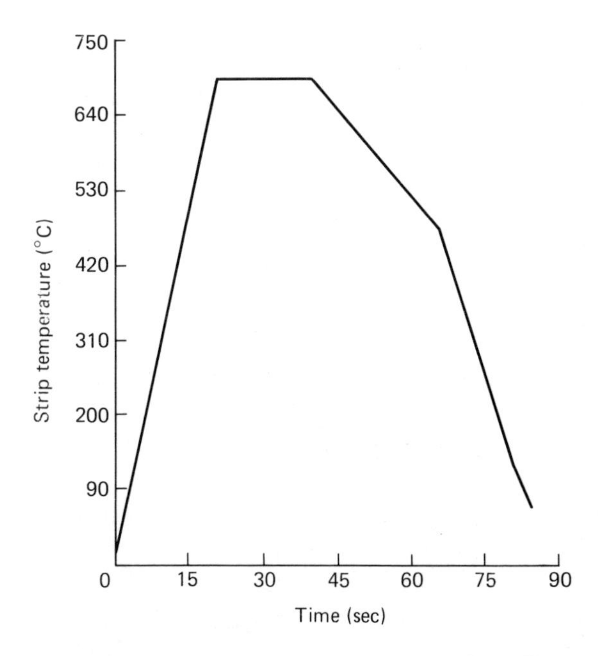

FIGURE 2.9.7 TEMPERATURE GRADIENT IN C.A. FURNACE

The heating zone consists of a large insulated muffle, fitted with vertical planes of radiant tubes (usually gas fired), between pairs of which the strip is fed vertically. A high heat imput and the thin strip section obtaining in tinplate practice enables the strip to be heated up to nearly 700°C, i.e. just below the critical range, in about 20 seconds. Fig. 2.9.6 shows that this zone accommodates eight loops (passes). Very close temperature control within the whole section is vital, and is ensured by fitting a number of thermocouples throughout, together with means for local adjustment to heat imput. The soaking zone chamber is similar in structure, and is fitted with electrical resistance heating units to maintain an even temperature precisely throughout the chamber. The time within this zone is also about 20 seconds, and it will contain about eight vertical strands of strip.

The slow cooling insulated chamber ensures steady and uniform cooling down to less than 500°C by means of tubes through which cold air is passed.

The final fast cooling chamber can be operated by a number of means for rapid removal of residual heat in the strip, which emerges from the protective atmosphere at a temperature of under 120°C, when there is no danger of a visible oxide film being formed on its surface. When sufficiently cooled it passes through carefully controlled tension reels to avoid plastic deformation, and is finally recoiled.

Modern CA furnaces are of sophisicated design, which will vary appreciably in their detail, see Price[41] and Johnson.[42] The four zones call for very precise control of heat input, coupled with temperature measurement which ensures a high standard of consistency; the lack of a direct measure of the degree of heat treatment available on the line, coupled with the effect of strip thickness, has, over recent years, led to the development of in-line means of checking this. One monitoring method has been developed by Youngstown,[43] based on a unit containing a magnetising head and probes to measure the magnetic field retained in the strip; hardness values are indicated directly in HR30T units, Its operation was found to be independent of strip speed, and is only slightly affected by strip thickness. Development of a similar type in the United Kingdom has been described by Syke.[44] Monitoring devices of this type are particularly useful as they can prevent production of substantial amounts of plate outside temper specification, as samples cannot be obtained from the coil until after the completion of annealing.

Many lines will contain up to 400 m of strip in the looper alone, and the total length within the whole assembly can exceed 1000 m. The need for proper tracking of the strip and precise tension control are therefore paramount, particularly in view of the elevated temperatures over appreciable lengths. There is thus a lower limit to the strip gauge which can be processed, and it is down to about 0.18 mm (0.007 in). As the dwell time just below the critical range is not more than 20 seconds, the time is too short for coalescence and grain growth and the ferrite grain size inevitably is comparatively small (Fig. 2.9.4). This is ideal where the highest levels of stiffness and yield are desired, but rather too fine for optimum ductility in deep drawing processes. As will be seen in a later chapter dealing with plastic deformation and canmaking processes, an intermediate grain size is ofter desirable, as a better compromise will be obtained between a softer grade and tensile strength. A much larger grain size, although resulting in a lower hardness, will lack strength. At present, therefore, continuous annealing in general will be used for the higher temper grades; for the softer tempers, batch annealing is to be preferred.

Large lines will handle coils weighing up to 20 000 kg, and as they operate at quite high speeds, substantial tonnages of plate are processed. In many situations, these would be far too large, but smaller, slower strand annealers are used. They operate at speeds of up to 150 m/min and are usually designed to feed strip horizontally. In addition to constructional and operating advantages, their time/temperature cycles are more amenable to providing somewhat larger ferrite grain sizes for the softer temper grades. Many modern high-speed furnace developments, particularly in Japan, are capable of providing the conditions needed for producing wider temper ranges.

2.10 TEMPER ROLLING AND COIL PREPARATION

After annealing, the steel strip will be in a soft pliable state and will possess a rough "open" surface. In the case of double reduced plate, as described in that section, a second heavy cold rolling operation is used, and temper rolling is not required. But single cold reduced plate must be given a light cold rolling, termed temper rolling (or "skin passing") before proceeding to the final surface finishing treatment. Temper rolling is carried out for three essential reasons:

(1) to provide a smooth surface finish (or special finish, below);
(2) to give a flat sheet; and
(3) to achieve the desired mechanical properties according to the particular temper grade specified.

Any of the special surface finishes described in Chapter 3 can be formed in the last stand of the temper mill by appropriate treatment of its working roll surface — similar to that described under double reduction rolling.

A high standard of flatness is ensured by grinding a "crown" into the middle of the working rolls (as in cold reduction, section 2.8) and by careful control of strip tension and roll pressure; good basic rolling requirements are similar to those called for in cold reduction.

The extension applied to the material on temper rolling will vary, usually from less than 1% for the softest temper grade — this being generally known as "skin-pass" — increasing to a few per cent for the harder tempers; but the elongation aimed for will be varied according to the chemical composition of the steel and the annealing cycle employed. For "built-up" bodies (Chapter 4), in which the plate blank is rolled into a cylindrical form, or one of oval cross-section, the steel will need to be adequately temper rolled to ensure that any fluting in the profile during forming is eliminated (the alternative term panelling is sometimes used). This phenomenon is the result of uneven bending and sudden yielding, and is associated with the elongation observed immediately beyond the yield point in the tensile test stress-strain curve, and thus also with the formation of Lüderlines on the test specimen (and in deep drawing). The deformation characteristics of low carbon steel are discussed in some detail in section 5.2; as will be seen, the phenomenon of age-hardening also plays a part in this behaviour, as this may cause a partial return of the yield-point elongation (Fig. 2.10.1) if stored for long times and/or subjected to higher temperatures such as in tinning or several stoving treatments in lacquering processes. Age-hardening is related to nitrogen and carbon contents in supersaturated solution; if the steel is of the killed variety, it does not exhibit a yield-point elongation.

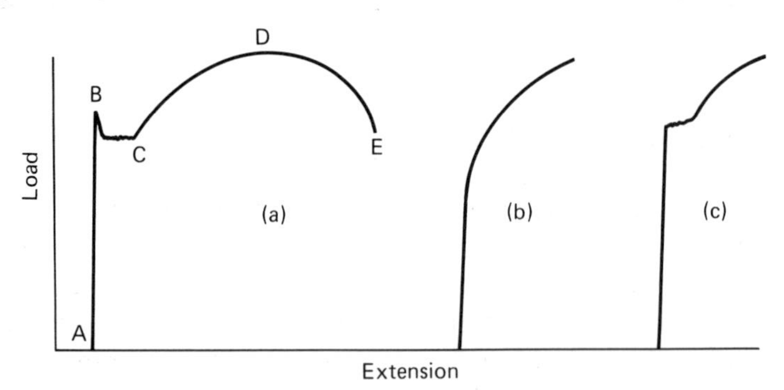

FIGURE 2.10.1 LOAD-EXTENSION CURVES FOR LOW CARBON STEEL: (a) ANNEALED; (b) ANNEALED AND SKIN-PASSED; (c) ANNEALED, SKIN-PASSED AND AGED

Temper rolling is normally carried out in four-high mills and, particularly for the harder temper grades requiring greater material extension, a two-stand four-high mill will, in general, be used.

Rolling speeds will often be in excess of 1500 m/min, and the mill will be equiped with sophisticated instrumentation to provide control of rolling loads and strip tension. Reduction in strip thickness is minimal, and a rolling lubricant is not used; in fact a high standard of cleanliness is required, as it is vital to avoid contamination of the surface, which may be difficult to remove prior to tinning (Fig. 2.10.2).

FIGURE 2.10.2 2-STAND TEMPER MILL

Coil preparation

This operation is aimed primarily at inspecting the surface quality of the strip and removing any faulty portions that may be present; this will include any off-gauge material detected by means of x-ray or nucleonic measuring units and pinhole detectors. Standard practice will include rolling to a width at least 12.5 mm (0.50 in) greater than the ordered dimension, and the line will normally include a pair of accurate rotary knives to trim each edge of the strip to within the tolerance imposed on the ordered width dimension. The line will thus usually include uncoilers and recoilers, cutting, rejection and welding units, in addition to thickness gauges, pinhole detectors, side trim rotary knives and precise mechanism for location on the strip. This combination will provide large coils, up to 20 000 kg in weight, for the final coating line, depending on the number of welds which have to be made after rejection of portions; in view of interruption to fast operation caused by the presence of welds, canmakers normally lay down a maximum of four in any one coil. Some lines will contain a burr masher to flatten the burr formed on edge slitting. Some units will operate at speeds up to 1500 m/min, but they will vary widely in speed and the type of equipment provided.

Occasionally the steel strip will be cut up into individual sheets prior to metallic coating, if they are required to be tincoated by the now seldom used hot-dipping method (described in the next section), or as uncoated "blackplate" sheets. As this shearing operation is generally carried out on coated plate, the operation is described in Fig. 2.12.

2.11 DOUBLE REDUCED PLATE

Conventional plate is virtually in a fully annealed state, as the extent of thickness reduction during temper rolling is very small; consequently its ultimate tensile strength is generally less than 500 N/mm^2.

As demands for stronger tinplate grew, to enable appreciable reductions in thickness to be made, many organisations developed the technique of double reduction, which involves a second cold reduction after process annealing, to increase the steel base strength. The effect of cold reduction on UTS is shown in Fig. 2.11.1. This increase in strength, to an order of 650 N/mm^2, allows a

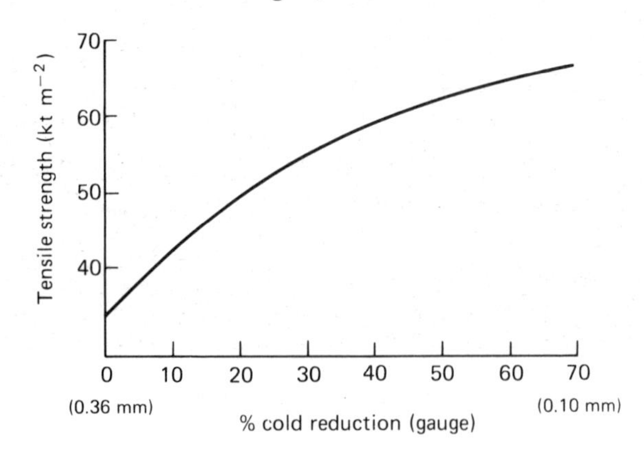

FIGURE 2.11.1 EFFECT OF % COLD REDUCTION ON THE TENSILE STRENGTH OF DR TINPLATE

component to be made from appreciably thinner plate without its suffering a loss of rigidity or having inferior deformation characteristics. The development has led to appreciable savings in material to many areas, and it is estimated that the usage of DR plate is as high as 25% in some areas.

The normal range of thickness reduction is 30-40%, but occasionally a higher level may be adopted. The bulk of DR plate is in the thickness range 0.16-0.18 mm; to reduce thickness down to this level, very heavy duty mills and sophisticated lubricating systems are required. They are usually two- or three-stand four-high tandem mills (Fig. 2.11.2). It is claimed that the last stand should only carry out a small part of the total thickness reduction required in order to achieve good strip shape, and many operators argue the merits of a three-stand mill. In some cases, single-stand reversing mills have been used. Rolling techniques are demanding and good lubrication systems are vital; these requirements and trials on a number of lubricant types are discussed by Vucich and Vitellas.[45]

FIGURE 2.11.2a THREE-STAND MILL

The process has now been in use for some years and the British Steel Corporation has prepared a bibliography on various aspects of the process for the period up to 1973.[45]

The various grades of DR plate commercially available and their properties are described in Chapter 3; but it should be mentioned here that the effect of a thickness reduction of the order of one-third increases strength at the expense of ductility. The plate properties are significantly more directional, and when it is to be used for forming cylindrical build-up bodies the rolling direction should be specified; the reasons for this are also discussed in Chapter 3. Avoiding split flanges is more critical than when using single reduced plate and an unusually thin area adjacent to the sheet edge can also be troublesome. The practice of using this type of plate has now become well established.

FIGURE 2.11.2b TWO-STAND MILL

2.12 METALLIC COATING

2.12.1 Hot dip tinning

The amount of plate coated with tin by this method is now small, and decreasing continually. Detailed accounts can be obtained from *The Technology of Tinplate*[19] and the *Making Shaping and Treatment of Steel*.[47] It consists of passing individual sheets, or narrow continuous strip in a few countries, through a bath or molten pure tin after appropriate surface treatments:

(i) The steel surface is first cleaned and any oxide present removed by a pickling process similar to that described earlier under hot rolling, and then thoroughly washed of all acid residues.

(ii) The sheet is fed through a flux layer of a solution of zinc chloride with some ammonium chloride, which floats on the tin.

(iii) As it passes through the molten tin, the sheet will be coated with the metal and an intermetallic compound having the formula $FeSn_2$ will be formed between the tin and the steel substrate; the tinning machine is illustrated in Fig. 2.12.1.

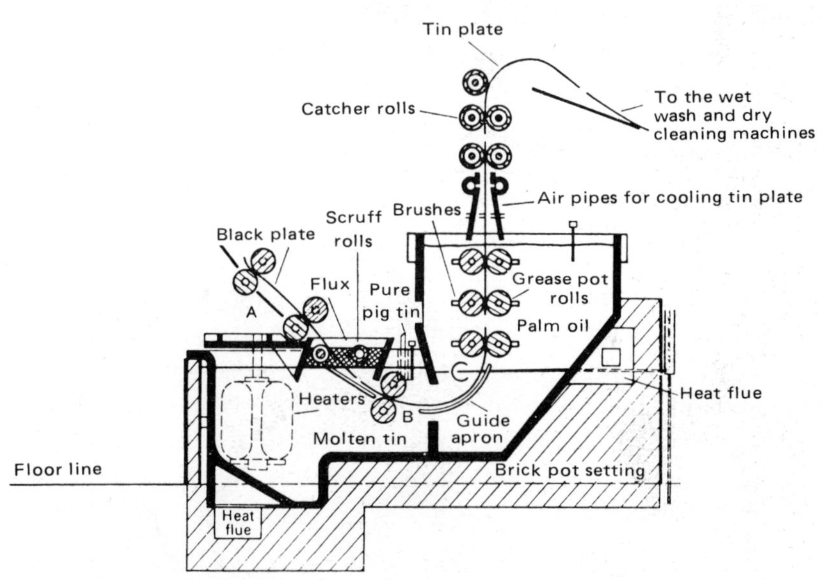

FIGURE 2.12.1 HOT DIP TINNING POT (SINGLE SWEEP) FOR SHEET STEEL

As the sheet emerges from the bath, it is taken up by a number of pairs of rolls and submerged in a bath of palm oil floating on the surface of the molten tin. This provides the means for controlling the amount of tin remaining on the sheet surface; palm oil is effective in preventing dewetting of the liquid tin, and the temperature is controlled so that the tin is on the point of solidifying shortly after it leaves the palm oil pot. A final cleaning operation to remove excess palm oil is carried out. It is important to avoid an excessive oil residue, as this will interfere with effective application and coverage of lacquers and organic coatings on to its surface, free of dewetting.

Although the process has undergone appreciable improvement over the centuries — a crude form was used in the fourteenth century — it suffers from some shortcomings in comparison with the modern electrolytic process described in the next section:

(i) the major disadvantage is that it cannot apply a fairly uniform coating lighter than 20 g/cm^2;
(ii) the coating is appreciably less uniform at much higher coating levels, it is slower in operation, and more expensive;
(iii) it cannot apply different coating weights on the two surfaces;
(iv) electrolytically tinned plate can be supplied in wide, large coils, a significant advantage in high-speed processing.

2.12.2 Electrolytic tinning

The process of depositing the tin by electroplating has been developed over the last 50 years into a very fast and accurate process, with sophisticated control systems incorporated throughout the line. It is capable of rapid change in the nominal coating weight applied, and of depositing different coating weights on the two surfaces.

The coating as deposited has a matt light grey appearance, i.e. it is not bright plated, but is readily converted into a lustrous bright appearance by momentarily raising its temperature above the melting point of tin and cooling rapidly. This is common to all of the electrolyte systems used commercially. There are four in common use, all of which have been developed into highly sophisticated processes; these are:

1. The acid stannous sulphate bath; this has been developed mainly by the United States Steel Corporation. It is generally known as the Ferrostan process, and is very widely used under licence.
2. The Halogen process; this was developed by E.I. du Pont de Nemour, Weirton Steel and Wean Engineering, and is probably the second most widely used.
3. The Fluoborate method was developed by Rasselstein in West Germany and is used also by other producers under licence.
4. The alkaline stannate bath; this was also developed some years ago, but is not so widely used nowadays, and is confined to the early lines installed.

Apart from the electrolyte and plating conditions used, there are many basic similarities between them. Gabe[20] quotes the following as typical compositions and plating conditions.

TABLE 2.12.1 (also Hoarse[19])

	Temp. °C	Cath.C.D. A/dm^2	CCE%	ANODE	NOTE
Acid sulphate					
30 g/l Sn^{2}+; 20 g/l free acid (PSA). Addn. agent 2-5 g/l	40-50	10-40	95	Sn	Tinplate
Alkaline stannate					
(i) 70 g/l K_2SnO_3; 27 g/l NaOH	80	4	90	Sn	
(ii) (Na) 45 g/l Sn^{IV}; 15 g/l free OH^-	85	5-8	60	Sn	Tinplate
(K) 150 g/l Sn^{IV}; 22 g/l free OH^-	90	25	80	Sn	
Halogen 75 g/l $SnCl_2$; 25 g/l NaF;50 g/l KF_2; 45 g/l NaCl; pH = 2.7	65	45			Additive required for tinplate

The "Ferrostan" process uses high current densities, and the heating effect is appreciable, so external cooling of the electrolyte is carried out as it is circulated continuously. A modern line will operate at up to 600 m/min (2000 ft/min). Very precise control of plated weight is supplemented by direct measurement using x- or γ-ray techniques. The complex line conditions are frequently computer controlled. In view of the high cost of tin, effective control of drag-out and rinsing liquor is practised. Plating is preceded by cleaning in a pickling and degreasing unit, both being electrolytic, and the strip is very thoroughly washed subsequently. After the plating stage, the coating is flow-melted, then passivated and finally lightly oiled. Flow melting consists of heating the strip to a temperature above the melting point of tin, say 260-270°C, followed by rapid quenching in good quality water. The temperature is raised and controlled by resistance or induction heating, or both. During this treatment, a small quantity of the tin-iron compound $FeSn_2$ is formed; its weight and structure will depend on the time interval and temperature reached, as well as on several other factors, such as the surface condition of the steel, the plating conditions and liquor residues on the plate surface. As will be described later, the structure and weight of this alloy layer plays an important role in several forms of corrosion behaviour, and sometimes also on soldering performance, especially of the more lightly plated grades. The amount of tin coverted to alloy is substantially lower than on hot-dipped tinplate, and seems to be virtually independent of the total tincoating weight. It is vital to maintain as low a weight as possible of compound layer, consistent with a compact continuous structure; this can be ensured by employing very rapid rates of heating, of the order of 150°C/sec, and by circulating the cooling water to prevent its overheating. This practice will also ensure no danger of contact of the tin with the carrier rolls, whilst still in a molten state. A good quality water is always used for quenching, e.g. steam condensate, so as to avoid the incidence of spots and stains on the surface, which can affect the lacquer adhesion and corrosion resistance of the plate, and detract from its appearance. Transition from matt to a brilliant surface is usually instantaneous, and any area not properly flow-melted will be readily detected.

Differentially coated electrolytic tinplate, the term used to describe plate carrying a lighter weight on one surface than the other, needs to be marked to identify the lighter from the heavier coating; this is normally done by applying, before melting, a solution which will prevent complete brightening along the agreed pattern. Most specifications recommend a pattern of parallel lines, the distance apart being related to the differential weight combination. It is more usual to apply the pattern on the more heavily coated surface, but by agreement they can be applied on the other. If properly applied, the lines have no effect on the plate performance, but it is essential that the specified line width is not exceeded. The solution used for this purpose is usually dilute sodium dichromate or sodium carbonate.

Passivation treatment

The naturally formed oxide layer on the surface of tin will readily grow in the atmosphere, especially when heated, e.g. when stoving applied organic coatings, to the point of being visible as a yellow stain; this can seriously affect the coating adhesion to the plate surface. (This growth also happens during long-term storage, especially in tropical conditions.) The strip is therefore next given a passivation treatment to render its surface more stable and resistant to the atmosphere, particularly at higher temperatures; the film, which is usually less than 1 μ thick, is also effective in preventing sulphide staining by some foods, and in improving lacquer coverage and adhesion. It also has a beneficial effect on solderability. This treatment is one of the most important stages, and many types of treatments have been evolved, e.g. in an aqueous solution of chromic acid, but overall an electrolytic treatment in a sodium dichromate electrolyte has gained widespread favour for applications involving the use of unlacquered and lacquered plate. The film is composed basically of chromium and chromium oxides and tin oxides; its varying properties appear to be dependent on the quantity and form of these basic constituents.[51] It is generally agreed that a better performance is achieved with meat, fish and soup products with a film having a relatively high chromium and a low tin content, whereas milk products require a film low in chromium and high in tin. Preferred treatments will be found to vary between packers and regions. Numerical classifications are used by one plate manufacturer to differentiate the various types. To determine the optimum conditions of treatment, especially for new lacquer types or new canmaking processes, requires very thorough testing.

After passivation, the plate is given a light oiling to help preserve it from attack, to assist in feeding sheets through machines, and the wetting of lacquer films. It is vital that the oil used is fully acceptable for use in food packaging, and the amounts must lie within narrow laid-down limits. Cotton seed oil had been used for many years, but was superseded by di-octylsebacate (DOS) some years ago, as this appeared to give a more stable performance on plate stored for several years in normal atmospheres, and also maintained lubricity of the plate surface to a higher level. (Friction tests have been developed to check this performance, and also several tests for measuring the film weight applied.[48]) Latterly acetyl tributyl citrate (ATBC) is also in use. The oil is usually applied by an electrostatic method, in which "atomised" sprays of the oil are directed to the vertically rising strip as it is being conveyed between sets of grids carrying a very high electrostatic potential relative to the strip, which is earthed. The powerful electrostatic field maintained within the banks of fine wire grids causes the minute droplets to be deposited uniformly on the plate surface. Close control of the oil spray jets and of the electrostatic field ensure an accurate and uniform film of oil; too little locally will contribute to rusting problems, while an excessive amount can give rise to lacquer dewetting. Oil film weights are generally in the range 5-10 mg/m^2; other levels for particular purposes can be obtained by agreement.

Diagrams of Ferrostan and Halogen electrolytic tinning lines are given in Fig. 2.12.2. The remainder of the line consists of several inspection and classification sections, prior to the strip being sheared into sheets or coiled, and packed for shipment. The checks carried out on these sections of the line include photoelectric cell units to detect pinholes through the steel base, thickness guages reject portions showing excessive variation, in addition to visual inspection for surface quality. As the strip speed on modern ET lines is often in excess of 500 m/min, visual inspection is not adequate to identify all surface defects; this has led to the development of many precise optical/electronic systems for more positive rejection of surface faults. Significant improvement has also been achieved in more precise detection and rejection of pinholes, particularly those located near the strip edge. Demands for narrower thickness tolerances to meet the requirements of the new canmaking methods has led to the development of more precise thickness measuring gauges, coupled with improved reject systems.

Many ET lines have included x-ray fluorescence gauges designed to give continuous readings of the tincoating weight being applied, including means for providing a print-out record; these give the total tin applied and the weight to each surface separately. The sensors can be stationary or traverse the full width of the strip. They ensure that the tincoating weight applied is more uniform and nearer the nominal weight that has hitherto been possible, and also avoid the transitional wander in weight which tended to occur on changing the line from one nominal weight to another. Descriptions of the apparatus and their performance have been given by US Steel[(49)] and by the BHP Steelworks at Port Kembla Australia.[(50)] The units in each case were developed by US Steel in collaboration with Nucleonic Data Systems.

Numerous improvements of these types have been made by many tinplate manufacturers, and as a consequence the overall quality of the Standard Grade tinplate (discussed in Chapter 3) has shown significant improvement in most areas.

These improved "on-line" procedures have been backed up by sophisticated quality control procedures, which provide for routine checks on physical, metallurgical and chemical characteristics. All those laid down in the relevant standards are checked regularly; these will be complemented by analytical tests on degreasing, pickling and tinning solutions for alkalinity/acidity, salts concentration, etc., and in the case of the tinning electrolyte, a Hull cell is used to test the plating characteristics of a sample. This involves determining the effective current density range of the sample; its use is necessary, as the additives included in the electrolyte cannot be checked by analytical methods. The precise nature of the quality control tests will vary considerably between plants; frequently quality assurance systems are set up jointly between plate manufacturers and canmakers. In addition to the specification tests described in Chapter 3, the numerous process tests are described in considerable detail in *The Technology of Tinplate*.[(19)]

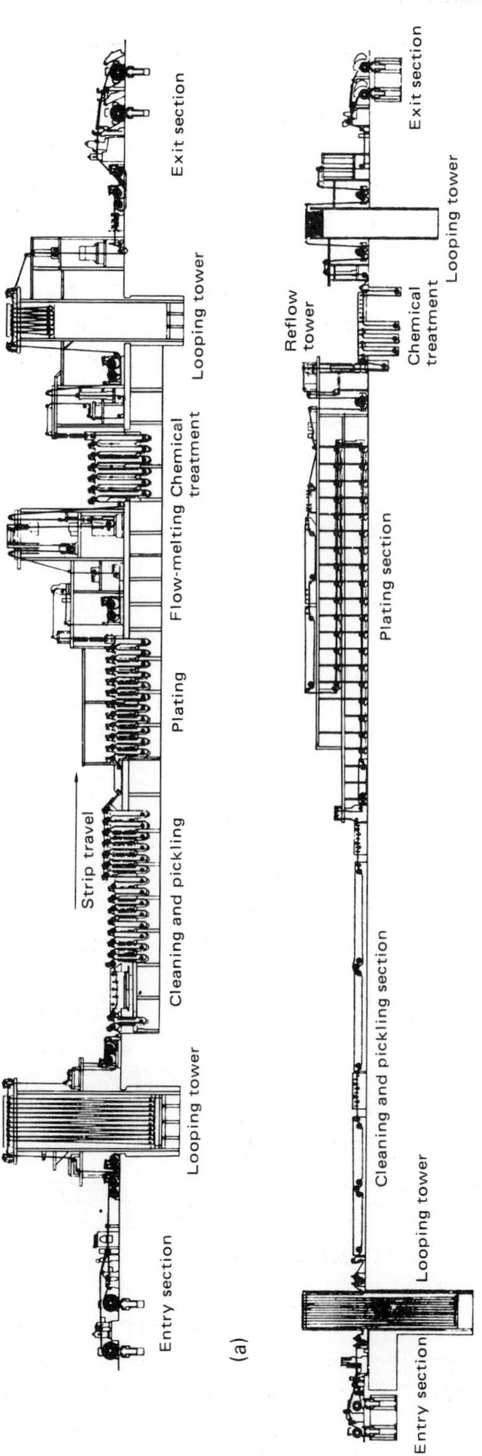

FIGURE 2.12.2 (a) FERROSTAN ELECTROTINNING LINE (b) HALOGEN ELECTROTINNING LINE

2.12.3 Alternative metallic coatings

Although uncoated steel sheet ("blackplate"), oiled or not, has been widely tested experimentally in many applications, it has not achieved significant usage, apart from periods of hostilities when tin was virtually unobtainable. This is because it has to be effectively protected against corrision on both surfaces by lacquering, and this has to be carried out almost immediately on receipt at the can-making plant, often very difficult to arrange. The protection given by the lacquer film is destroyed if it is marred, and the steel is vulnerable to severe corrosion attack, in the absence of any galvanic protection. Its appearance after lacquering also is poor.

Attempts have been made to improve its resistance to the effects of damage by applying light phosphate or chromate treatments, but without significant improvement.

The only type which has achieved significant success is that coated with metallic chromium. Its development was given considerable impetus during the 1960 decade when the price of tin was extremely variable and overall rose from £1000 to £1600 per tonne; on occasions it was also in short supply. At the time or writing it has increased to more than £9000 per tonne in the UK. Since that time the use of chromium coatings has increased significantly, but of course substantially less than tinplate.

During its early development the material was given the unfortunate title of Tin Free Steel, which is hardly appropriate. It is now generally known as (electrolytically coated) chromium/chromium oxide low carbon steel sheet, a reasonably precise description (Fig. 2.12.2).

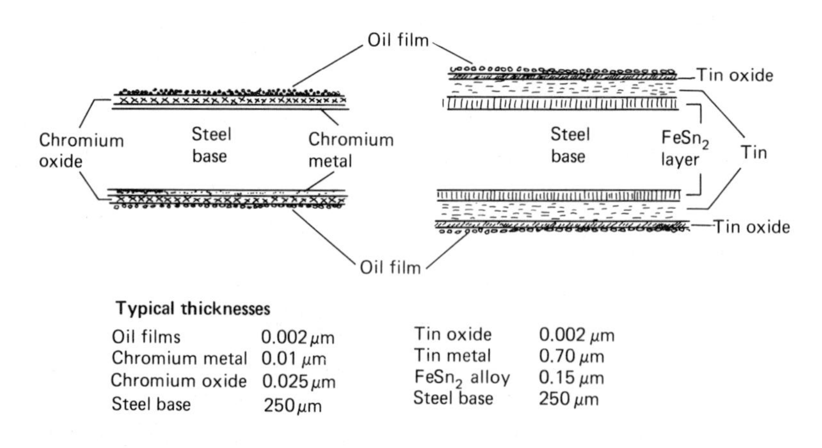

FIGURE 2.12.3 ALTERNATIVE COATINGS

The initial development work was carried out primarily in Japan; several versions have been described known as Hi Top, Cansuper, Supercoate (Japan), Ancrolyt (Germany) and Weirchrome (USA); all consist of a duplex coating of metallic chromium and chromium sesquioxide, and these can be applied in one stage, codepositing chromium metal and oxide, or two stages, in which the metallic chromium is deposited and then subjected to oxidation treatment. Smith[52] has given an account of the two forms of process. Some of these are being used under licence by several other manufacturers.

These coats give good lacquer adhesion, and have a good appearance. Lacquering on both surfaces is essential to avoid serious scratching of machine parts; none can be machine soldered, cementing of seams is possible and modern methods of welding are possible with certain coating weights.

Surveys have shown that the coating weights used, in terms of metallic chromium to chromium oxide ratio, as well as amount of metallic chromium, vary considerably between manufacturers and areas. The ideal seems to be in the ranges

0.7 – 1.5 mg/dm^2	Cr metal
0.3 – 0.6 mg/dm^2	trivalent Cr (present as oxide)

Thus the total coating weight approximate to 1.5 mg/dm^2 in comparison with 56 mg/dm^2 (i.e. 5.6 gm^2), the lowest normal grade of electrolytic tinplate; this indicates that the metal cost is substantially higher in the case of electrolytic tinplate. But a cost comparison on this basis can be misleading; many other factors need to be taken into account, such as capital expenditure, if the demand for the plate is sufficient to warrant a line solely devoted to this product. It is more usual in fact to convert smaller ET lines to produce both types of plate, as dictated by the demand, thus requiring additional capital expenditure, together with slower production speeds and greater down-time due to change-overs. In general, additional lacquering is also likely to be involved, but the final analysis shows a significant cost saving, especially when the tinplate container is also lacquered on both surfaces.

There is considerable similarity between an electrotinning and a Cr/CrO_x line, the only essential differences being that in the latter case flow melting and chemical passivation are not involved. Gabe[20] describes the process as cathodic deposition in a dulute chromium plating electrolyte — e.g. 50 g/l CrO_3 and 0.50 g/l H_2SO_4 at a temperature in the range 50-70°C and current densities up to 100 A/dm^2 for a line speed of 350 m/min.

The material can be used for a number of products, especially the more inert types; and processed foods can be included, particularly for the end component or a body not incorporating a side seam.

There is also interest in electrodeposited nickel as a possible alternative coating to tin. It is reported that extensive trials are being carried out in the United States.

2.13 CUTTING INTO SHEET AND PACKAGING

Tinplate has been delivered in sheet form since its early manufacture, and the sizes have increased over the centuries to the current average of about 950 mm × 800 mm (37.5 in × 31.5 in). The longest dimension can be in some cases as much as 1000 mm. Optimum sheet size is related to the component to be cut from it, so that wastage is at a minimum.

Plate in sheet form is packed normally on to a wooden platform ("stillage") to a convenient height and total weight; this will usually vary between 1000 and 2000 kg, according to sheet thickness and handling equipment available. The number of sheets in the bulk package will normally be a multiple of 100. Packing materials will be carefully selected to ensure that they are free from harmful materials, and the whole package is carefully wrapped and assembled to avoid the effects of moisture on the plate; when transported to or through areas of high humidity, the packages will be effectively sealed.

To improve on plate utilisation, coils approaching 10 000 m in length may be used; special packaging methods have been developed for this purpose, particularly to avoid physical damage.

REFERENCES

1. *Metal Bulletin*, 16 March 1982, 33.
2. *Direct Reduction of Iron Ore: a Bibliographical Survey by the European Coal and Steel Commity*, 1976 (in German). English translation: Metals Society, London.
3. *Metal Bulletin*, 27 April 1982, December, 23.
4. *Metal Bulletin Monthly*, December 1981, 49.
5. GOETTE and SANZENBACHER, *Electric Fce. Proc.*, 38, 22.
6. R.L. STEPHENSON, *Direct Reduction Update*, p.40, I & SM, 1982.
7. Rolling Steel Sheet from Powder, *MPR*, October 1981, 498.
8. G.J. McMANUS, *Iron Age*, June 1981, MP5.
9. 75th BISRA Steelmaking Conference, Scarborough, 1970.

10. BCS/BISPA Conference, Scarborough, 1973.
11. N. WILTSHIRE, *Proc. First International Tinplate Conference*, p.40, ITRI, London, 1976.
12. D.H. HOUSEMAN, Refractories and the Newer Steelmaking Processes, *Bull. Fuel and Metals*, (1978), 101.
13. C. HOLDEN, *Iron & Steel Inst.* 207 (1969) 806.
14. Z. YAMAMOTO *et al.* RH Reactor — Recent Developments of RH Vacuum Treatment at Nippon Steel Corp. *Proc. Third IISC, ASM and AIMI Conference*, 615 Chicago, 1978.
15. T. KOHNO *et al.* Composition Adjustment by New Methods, *Steelmaking Proc. ISS – AIME*, 62, (1979) 125.
16. N.E. MOORE, A.O'CONNOR & G.L. THOMAS, *Iron & Steel Inst.*, 118 (1978) 600.
17. J.M. KEEFE, I. EARNSHAW and P.A. SCHOFIELD, *Flat Rolling: a Comparison of Rolling Mill Types*, Metals Society, 1979. Also *Ironmaking and Steelmaking*, No 4, 1979.
18. International Conference on Steel Rolling, Iron & Steel Inst. of Japan, 1980.
19. W.E. HOARE *et al. The Technology of Tinplate*. Edward Arnold, London, 1965.
20. D.R. GABE, *Principles of Metal Surface Treatment and Protection*, 2nd edition, Pergamon, Oxford, 1978.
21. J.S. BOOTH, The Four-high Reversing Mill for Strip. Sheet and Tinplate, *Metals & Materials*, November 1977, 39.
22. I. EARNSHAW, Cold Rolling of Sheet and Strip Steel, *Metals Technology*, July/August 1975, 339.
23. T. FURUYA *et al.* New Design Six-high Mill (the HC mill) Solves Shape Problems, *Iron & Steel* Eng, 56, (8) 1979, 40.
24. G.J. MCMANUS, Japanese Six-high Mill Grabs Attention in U.S., *Iron Age*, August 1980, MP21.
25. D. COLLINSON, Shape Measurement and Control of Rolled Strip, *Engineering*, July 1979, 936.
26. *I.H.I. Shape Meter* (pamphlet), Ishikawajima-Harima Heavy Industries Co. Ltd., Tokyo, Japan.
27. D.W. DENDLE, A Review of Automatic Control Systems for Cold Tandem Mills, *Steel Times International*, March 1979, 78.
28. D. HOBSON, Automatic Gauge Control for Mills, *Engineering*, June 1978, 571.
29. G.F. BRYANT (Editor) *Automation of Tandem Mills: a Comprehensive Review*, Iron & Steel Inst., London, 1973.
30. R.E. KALMAN, Phase-plane Analysis of Automatic Control Systems with Non-linear Gain Elements., *AIEE. Trans.*, **73** (1955) 383.
31. W.D. KING and R.M. SILLS, New Approaches to Cold Mill Gauge Control, *Iron & Steel* Eng., May 1973.
32. R. KOPP and H. WIEGELS, Kaltwalzmodell Bander Bleche, *Rohte*, 2, 1978.

33. H.J. TER MAAT, Controller Design for Time Lag Systems with On-linear Elements. *Journal A,* **20**, (1979) 203.
34. H.J. TER MAAT, The Renovation and Automation of a Tandem Cold-rolling Mill. *Automatica* **78**, (1) (1982) 63.
35. Weirton Brings in Continuous-Continuous Steel Rolling via No. 9 Tandem Mill, *33 Magazine*, September 1976, 44.
36. N. MORRIS *et al*, Review of Cold-rolling Mills. Paper 9, Met. Soc. Int. Conf., Cardiff, 1978.
37. H. SCHMITZ *et al.* Current Development Trends in Cold Mills, *Stahl Eisen,* **101**, (13-14) 1981, 123 (in German).
38. L.W. DUNN, Cold Mill Automation, *Iron & Steel* Eng., February 1975, 42.
39. B.C. BRADLEY, M.O. SMITH and H.N. COX, US Steel Fairfield Works Six-stand Computer Controlled Cold Mill, *Iron & Steel Eng.*, May 1970, 57.
40. D. McLEAN *Trans. Metall. Soc. AIME,* **242** (1968), 1193.
41. W.O.W. PRICE and E.H. VAUGHAN, *Conventional Strand Annealing*, Iron & Steel Inst. Special Report No. 79.
42. R. JOHNSON and J. DUNNING, *Continuous Annealing of Tinplate*, p. 126 Australian Iron & Steel Ltd., Annual Conference, Port Kembla, 1966.
43. Continuous Strip-hardness Readings from Magnetic Noncontacting Gage, *33 Magazine*, October 1976, 41.
44. G. SYKE and I. MURRAY, Measurement of Hardness in a Continuous Annealer: Non-destructive Examination applied to Process Control in the Steel Industry. Conference, University College of Swansea, 1967.
45. M.G. VUCICH and M.X. VITELLAS, Emulsion Control and Oil Recovery on the Lubricating System of Double-reduction Mills, *Iron & Steel Eng.*, December 1976, 29.
46. J. SLOAN *Bibliography on Double Reduced Tinplate*, British Steel Corporation, Strip Mills Division, Port Talbot, 1973.
47. *The Making, Shaping and Treatment of Steel*, US Steel Corporation (USA), 1971.
48. *A Guide to Tinplate*, ITRI, London, 1983.
49. J.R. GIBSON, Tin Coating Weight Gage at US Steel's Pittsburgh Works, *Iron & Steel Eng.*, January 1977, 51.
50. D. SALM, N.D. WILTSHIRE and D. KAAN, The Continuous Monitoring of the Tin Coating Mass on Tinplate, *BHP Tech. Bull.* **21** (2) (1977), 31.
51. J.P. COAD, B.W. MOTT, G.D. HARDEN and J.F. WALPOLE, Nature of Chromium on Passivated Tinplate, *Brit. Corros.* **11** (4), 1976, 219.
52. E.J. SMITH, *Iron & Steel Eng.*, **44** (7), (1967), 125.
53. S. YONEZAKI, *Metal Fin. J.* **16** (1970), 4.

FURTHER READING

PEACY and DAVENPORT, *The Iron Blast Furnace,* Pergamon Press, Oxford, 1979.

The Making, Shaping and Treating of Steel, US Steel Corporation (USA), 1971.

W.K. LU (Editor) *Blast Furnace Ironmaking,* 1977, and *Blast Furnace Ironmaking,* 1978, McMaster University, Hamilton, Canada.

STRASSBURGER (Editor) *Blast Furnace – Theory and Practice,* Gordon & Breach, New York, 1969.

C. MOORE and R.I. MARCHALL, *Modern Steelmaking Methods,* Institute of Metallurgists, 1980, Monograph No. 6.

Modern Steelmaking Practice, Pergamon Press, Oxford, 1983.

Principles of Rolling

Elements of Rolling Practice, 2nd edition, The United Steel Co. Ltd., Sheffield, 1963.

E.C. LARKE, *The Rolling of Strip, Sheet and Plate,* 2nd edition, Chapman & Hall, London, 1963.

The Making, Shaping and Treating of Steel, 9th edition, US Steel Corporation, Pittsburgh, 1970.

A.I. TSELIKOV and V.V. SMIRNOV, *Rolling Mills,* Pergamon Press, New York, 1965.

Z. WUSATOWSKI, *Fundamentals of Rolling,* Pergamon Press, New York, 1969.

GEO E. DIETER, *Mechanical Metallurgy,* McGraw-Hill, New York, 1976.

J.N. HARRIS, *Mechanical Working of Metals,* Pergamon Press, Oxford, 1983.

Chapter 3

Properties of Tinplate and Testing Procedures

3.1 INTRODUCTION

The properties and characteristics of tinplate cover many aspects, including appearance and surface faults, mechanical properties, tincoating thickness and structure, gauge, dimensions and shape. As generally supplied for the major range of industrial uses, tinplate has a bright reflective surface; but some applications require that the coating shall be "as deposited", which gives it a matt appearance, light grey in colour.

The manufacturer will apply a detailed quality control test procedure, from primary steelmaking through to the final stage of classification and inspection, prior to packaging. In addition, the purchaser will frequently subject a consignment of plate to a number of tests to ensure, as far as possible, that the material is suitable for the end use. Some of these will duplicate those already carried out by the manufacturer; others will compliment them. Of course this testing cannot be final, and the only absolute test is to subject the plate to the particular process. Most of tests used by the purchaser for this purpose are relatively straightforward and many will be carried out at the manufacturer's plant before the plate is shipped. They will cover visible surface defects, mechanical properties, amount of tin or other metallic coating, sheet dimensions, gauge and shape. When the plate is supplied in coil form, most of the testing will be carried out after the purchaser has cut it into sheet. The procedures used during initial inspection are usually those given in published standards and are described in this chapter.

As with most manufactured products, the characteristics of tinplate and its variants will vary from batch to batch, within a batch, and indeed from sheet to sheet. In order to obtain a reasonably accurate assessment as to whether the plate conforms with a specification, it is imperative that the sample is selected properly according to agreed procedures. These are described in detail in all modern standards (a list of all relative standards is given at the end of the chapter). Briefly, a number of bulk packages are selected from the consignment, and the number of sheets required is taken at random from each of the bulks; the number of sheets needed will vary according to the feature being tested and these are quoted in the relevant section. The method of taking specimans from each sheet for testing mechanical properties, amount of metallic coating, and gauge is shown in Fig. 3.1.1.

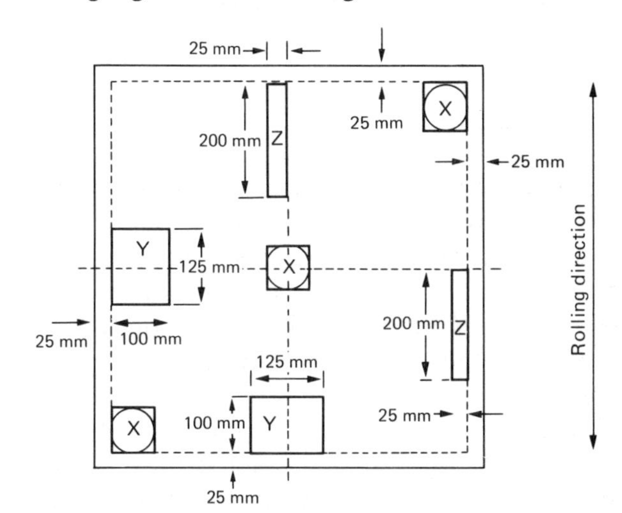

X – specimens for tincoating mass test
Y – specimens for hardness tests and determination of local thickness within a sheet
Z – specimens of tensile test or spring-back test

FIGURE 3.1.1 POSITION OF TEST SPECIMENS

On occasions, the purchaser may decide that the considerable effort involved in comprehensive sampling and testing is not justified (a bulk package will weigh up to 2000 kg). For example, the process may have been long established, or the plate has been supplied by one manufacturer for that purpose over a long period. In such cases only a limited inspection may be made, or possibly none at all, prior to use. There may be other occasions when the purchaser considers that a more detailed examination than that given in a standard is prudent; this could be done in the case of a new complex development. It must be emphasised that proper sampling must be carried out to judge conformation with a standard.

There are many other specialised tests available which relate to specific modern canmaking techniques, corrosion behaviour, or lacquering and printing performance; these generally require considerable expertise and sophisticated apparatus, and are often time-consuming. These are discussed later in the relevant chapter on canmaking or performance aspect.

In addition to these tests, which are of primary concern to the the canmaker, the packers will carry out an inspection of incoming batches of cans, usually concentrating on dimensional and visual aspects of quality; reliance is often placed on the can manufacturers to test thoroughly the basic quality of the plate. Plate thickness, seam dimensions, quantity of tin, and lacquer coverage are of primary importance to the packer, but for critical applications his testing may well extend to the more specialised plate parameters referred to in the previous paragraph.

3.2 SURFACE APPEARANCE AND GRADING

The final finish on tinplate is influenced by the nature of the steel surface prior to tinning, by the weight of tin deposited, and on whether the tin has been melted ("flow brightened") after deposition. As described in Chapter 2, the four recognised commercial finishes of single reduced tinplate are bright, stone, matt and silver; those of double reduced plate are more restricted.

Tinplate can exhibit a number of imperfections or defects which are readily visible to the unaided eye, but their appearance will vary according to the type of finish applied. In addition to their effect on appearance, these defects can also have an influence on solderability, corrosion behaviour and quality of coverage by lacquers and printing inks.

Surface quality is visually inspected before cutting the strip into sheets or recoiling. The standards of inspection inevitably are highly subjective, and neither the defects nor the standards to be adopted can be defined with any degree of precision. Clearly this operation requires highly trained inspectors who have gained a considerable experience, preferably of both the effects of plate manufacturing factors and user requirements. There are very many types of defects, and they can arise from many sources, as indicated by the examples given below:

Faults in	Examples of defects arising
Steel base	Non-metallic inclusions and laminations.
Rolling	Pinholes, ridges, oil spots.
Annealing	Stains and spots, burnt areas.

Tinning	Sparsely tinned areas, other tinning defects, incomplete flow brightening.

Electrolytic tinplate is classified into several grades: first quality, free from defects visible to the unaided eye, is termed Standard Grade. This is the product of the highly developed inspection and classification procedures which are applied at the end of the tinning line.

In some countries Second Grade is offered commercially; this plate will exhibit some surface imperfections, tinning defects and other faults to a minor degree, and its use will need to be considered with care. In carrying out this classification, the line inspector is able to trigger rejection of portions of the strip on which faults are detected, by simply pressing a button. This causes the sheets cut from that portion to fall into a separate reject classification pile situated beyond the shear. Although the strip speed can be up to 365 m per minute, he is able, with the aid of effective lighting conditions, to detect all significant imperfections. Modern electrotinning line speeds range up to 600 m/min, but above 365 m/min the physical performance of the sheet shear is inadequate, and it is the practice in such cases to recoil the strip and to cut up on a separate, slower, classification/shearing line. The underside of the strip is examined at the same time by means of a mirror. As the inspectors aim to meet fully the quality standard required, inevitably in an operation of this type some sheets of Standard Grade will be deposited into the reject pile: so it is the usual practice to re-examine meticulously by hand all sheets deflected into that pile, on a separate assorting table (as was done with the now obsolescent hot-dipped tinplate).

Additional reject piles accumulate the sheets found to be above or below gauge limits, or to have pinholes through them.

When strip is recoiled for delivery to purchasers in that form, a record of the incidence and location of any defects is retained by means of punched or magnetic tape, to provide a complete inspection history for the users during cutting up or on direct feeding into the press.

Hot-dipped tinplate is classified in a similar manner: as mentioned in Chapter 2, this historical method of tincoating is now obsolescent, and total world production of this type is believed to be less than 5% of total tinplate production. Hot-dipped tinplate, free from visible defects, is referred to as First Grade or Primes, and the second quality, having visible defects to a moderate degree, as

Second Grade or Seconds. The term Standard Grade for hot-dipped plate is different from and should not be confused with that used for electrolytic plate; it describes tinplate containing both First Grade and Second Grade unsegregated, but from which all other material exhibiting defects to a greater degree has been eliminated. Other terms used are:

"Menders" — plate showing defects which can be rectified by retinning.

"Wasters" — a Third Grade hot-dipped plate showing defects to a greater extent than is acceptable as Second Grade, and which can only be used for uncritical application.

Other metallic coatings and blackplate are also line-inspected visually for general appearance and incidence of faults; many of the types of defects appearing on electrolytic tinplate will also occur on these plates. As with tinplate, it is not possible to describe precisely the extent of visible defects permissible on these variants. Of course the basic appearance of blackplate and chromium/chromium oxide coated plate is different from tinplate, both being darker and less reflective.

To verify the grading of surface quality, most specifications stipulate sample sheets amounting to 1% of the total in each bulk package selected, of course to the nearest whole number; should disagreement occur, the retest must be made on 5% of the sheets in each bulk taken.

The four surface finishes described earlier are produced partly by employing various forms of roughened roll surfaces in the last stage of cold rolling. Surface microtopography and degree of roughness are of some importance, but it is generally accepted that there is no method yet available which will adequately characterise all the parameters contributing to surface quality; work is actively being carried out in many organisations to develop fully realistic methods. In the meantime, use is made of profile-measuring instruments to give some measure of the plate surface. The one most favoured at present is probably the Talysurf profile-measuring instrument; this employs a lighly loaded stylus to traverse the plate surface, and the fluctuating voltage generated by its movement is reproduced on a meter calibrated directly in microns or micro-inches either as CLA (centre line average) or RMS (root mean square). The signal produced can also be used to provide a trace of the surface profile. A range of magnifications is available. There are other instruments similar in principle also in use. One Euronorm (49-72), has been published to define a method for measuring the surface roughness of steel sheet; this also is considered to be inadequate for our purpose. In practice roughness can vary over the four grades from 0.12 μ to 2.5 μ (5 to 100 ins).

3.3 MECHANICAL PROPERTIES — GENERAL

Single reduced plate, the type which has been in use since the cold-reduction process was first developed about 50 years ago, is subjected to a light rolling only ("temper rolling") after the annealing treatment. Double reduced plate, which was evolved some 30 years later with the primary object of reducing raw material cost, is given a second major cold reduction after annealing. This modification has the effect of increasing appreciably the strength or stiffness of the plate, but of course at the expense of ductility. In view of its higher strength, the thickness range in which this grade is used commercially is significantly lower than that of single reduced plate. The method used for describing its mechanical properties is also different.

The grade of plate, the specifications describing their respective mechanical properties, and the test methods used to verify conformity with those specified are common to all plate variants — tinplate, blackplate and chromium/chromium oxide coated plate.

3.3.1 Single reduced plate

The term "temper" was adopted many years ago to describe a combination of interrelated mechanical properties such as hardness, ductility, springiness within tinplate technology, and is well known and understood within the realms of this industry. It is, of course, used in a very different sense from that in the technology of heat-treatable steels.

There is no single test available which can measure all the factors which contribute to the fabrication characteristics of tinplate materials. Experience over the years has shown that the Rockwell 30T hardness value (HR30T) is the most useful single test available, and this does serve as a useful guide to the mechanical properties of tinplate and its variants. A system of temper classification based on HR30T test values has thus been developed, and is given in Table 3.3.1; T50 represents the most ductile and soft grade, while T70 exhibits the highest strength and rigidity in the range and is the least ductile.

TABLE 3.3.1 TINPLATE TEMPER DESIGNATIONS: SINGLE REDUCED

Temper classification		HR 30T Hardness aim				
Current	Former	Mean	Max. deviation of sample av.	Approx UTS N/mm^2	Formability	Typical Usage
T50	T1	52 Max.		330*	Extra deep drawing	Normally stablised steel. Deep drawn parts.
T52	T2	52	+4 −4	350*	Deep drawing	
T55	—	55	+4 −3		General purpose	
T57	T3	57	+4 −3	370*	General purpose	For ends and round can bodies, Non-fluting.
T61	T4	61	+4 −4	415*	General, increased stiffness	For stiff ends and bodies, crown corks, shallow stampings.
T65	T5	65	+4 −4	450*	Resists buckling	For stiff ends and bodies.
T70	T6	70	+3 −4	530*	Very stiff	Beer and carbonated beverage can ends.

The Rockwell Superficial Hardness Tester (Fig. 3.3.1) employs an indenter in the form of a hard steel ball, manufactured with high precision, and normally a major load of 30 kgf, but major load of 15 kgf is used in the circumstances which are described below. The preliminary minor load is invariably 3 kgf. The anvil used must be the type fitted with a diamond spot centre, so as to standardise anvil conditions and minimise the indentation appearing on the under surface of the test specimen. In principle, the depth of permanent indentation — that is, after the major load has been removed — into the plate is measured; a sensitive gauge is incorporated in the instrument for this purpose. This gauge is calibrated "in reverse", so that the softer the material the greater the indentation, and thus the lower the reading obtained. It has a scale calibrated arbitrarily up to 100 units. The principle of the test is illustrated in Fig. 3.3.2.

The relationship between HR30T values and fabrication characteristics is valid only for the particular combination of factors normally obtaining in established tinplate manufacturing practice. Any appreciable departure will significantly affect the relationship, and therefore weaken the value of the hardness indicator. Indeed the HR30T value will vary with the normal variation in tinplate manufacturing practise; particularly with steel composition and casting conditions, hot rolling (rolling temperature and finishing temperature) cold rolling (extent of thickness reduction and type of mill), annealing (especially between batch and continuous annealing) and temper rolling. The manufacturer will aim at the mid-point of the hardness range of the selected temper grade shown in the table.

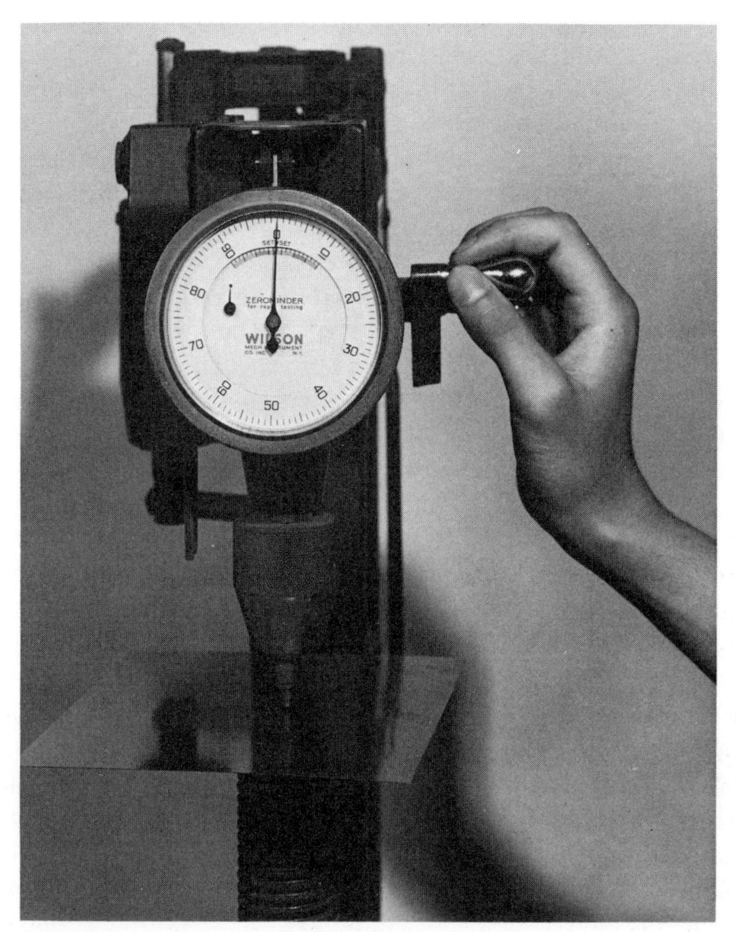

FIGURE 3.3.1 'SUPERFICIAL' ROCKWELL HARDNESS TESTING MACHINE

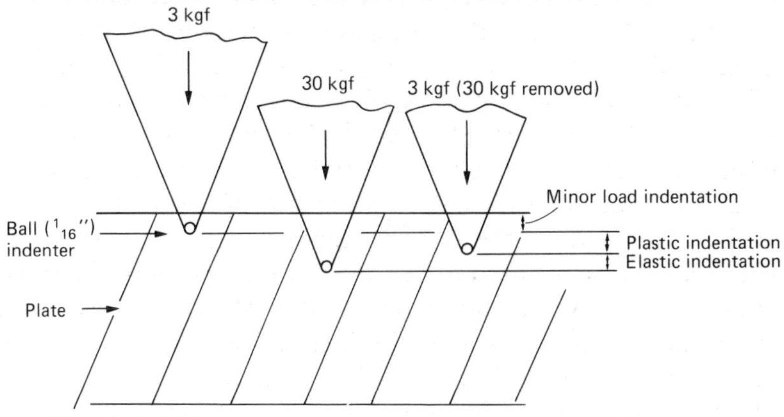

FIGURE 3.3.2 PRINCIPLE OF ROCKWELL HARDNESS (NOT TO SCALE)

Normally the canmaker will specify the most suitable temper grade for his particular operation and, if appropriate, whether batch or continuously annealed. Lower tempers will be selected for deep drawing operations to ensure that the plate has an adequate degree of ductility to withstand the deformation involved. Where appreciable intrinsic ductility is not required, a higher temper is chosen, as its higher strength will provide a better balance between high resistance of the container against damage due to rough handling and severe transport conditions and sufficient formability for the process. If there is doubt, the appropriate grade should be discussed with the manufacturers. In some cases the type of steel (e.g. killed or rimming) to be used will also need to be considered. The location of the test specimens is shown in Fig. 3.1.1 They should be 25 mm to 40 mm each way. Normally two sample sheets will be taken from each bulk package selected. It is essential that the tincoating is removed before carrying out the test, to avoid spurious results due to the softer tin metal; this error will vary according to the thickness of tin on the plate surface. The tin is readily removed by means of Clarke's detinning solution or any other established method (see appendix). Similarly, any other metallic coating must be removed before testing.

The "anvil effect" can significantly affect the recorded value at gauges toward the upper or lower limits of the range. The hardness values quoted in the table apply to plate in the middle of the thickness range, about 0.25 to 0.30 mm. Experience shows that material of the same metallurgical quality at a gauge of 0.22 mm is usually one HR30T unit higher, and at a gauge of say 0.43 mm would be one unit lower. At thicknesses below 0.20 mm on the corresponding value may be two units lower. It is fairly general practice when testing plate thinner than 0.20 mm to use the 15 kgf major load; the test values are then converted to HR30T using the conversion scale given in Table 3.3.2. Some laboratories adopt the practice of applying a correction factor for gauges outside the range of 0.25 – 0.30 mm, to compensate for this effect. This will not be more than a few units at most, and it is debatable whether any greater precision is achieved by this practice, in view of all the factors — processing conditions, age-hardening, etc. which can contribute to the value recorded. The test does not simulate closely any fabrication process, but it does, within limits, provide a useful guide to general properties.

TABLE 3.3.2. CONVERSION OF HR15T RESULTS TO HR30T VALUES

HR15T values	Equivalent HR30T value	HR15T values	Equivalent HR30T value
93.0	82.0	84.5	66.0
92.5	81.5	84.0	65.0
92.0	80.5	83.5	63.5
91.5	79.0	83.0	62.5
91.0	78.0	82.5	61.5
90.5	77.5	82.0	60.5
90.0	76.0	81.5	59.5
89.5	75.5	81.0	58.5
89.0	74.5	80.5	57.0
88.5	74.0	80.0	56.0
88.0	73.0	79.5	55.0
87.5	72.0	79.0	54.0
87.0	71.0	78.5	53.0
86.5	70.0	78.0	51.5
86.0	69.0	77.5	51.0
85.5	68.0	77.0	49.5
85.0	67.0		

Hardness evaluation is complicated by possible age-hardening effects during normal storage; for example, uncoated steel base (blackplate) could be up to four units lower than its corresponding finished tinplate. Age-hardening is not necessarily complete after electrotinning (even with flow-brightening), but will reach equilibrium after the stoving treatments involved in lacquering and printing processes, or after long storage. Effects of age-hardening are complicated, being dependent on many factors, such as carbon and nitrogen contents of the steel, annealing conditions, storage temperature and time, etc. The effects of incomplete age-hardening can be eliminated by immersing the test specimen in molten tin for 20 sec. at a temperature of 340°C, or placing it in an oven at 210°C for not less than 20 min. This treatment is recommended for non-flow melted tinplate, chromium/chromium oxide coated steel and blackplate.

It is also claimed by some workers that the hardness value recorded on the rougher grades of plate will be lower by one or two units than on comparable smooth plate, and the practice of smoothing the steel surface after removing any metallic coating, carefully with fine emery paper, is sometimes adopted.

It is strongly recommended that the practice of frequently checking the performance of the machine by means of the calibrated test blocks is rigorously followed. In addition, comparison of two or more machines, e.g. between manufacturer's and user's, is to be recommended. Preparation of test specimens for this purpose has to be done very carefully; an appropriate method consists of cutting up the central area — i.e. the more uniform area — of a sheet of flow-brightened tinplate into specimens of the required size and to randomise them into the number of groups required, using a table of random numbers. A further precaution is to fully age-harden them by oven heat treatment.

Other Tests Sometimes Applied

Tensile test. This is used for evaluating double reduced plate. Some workers contend that the relationship between the fabrication characteristics of single reduced plate and tensile yield strength is equal to that with the Rockwell HR30T. This test is not widely used, however, because of the additional skill and time needed to carry out accurate tensile testing of thin sheet. When some doubt exists regarding plate suitability, some workers will take into consideration both HR30T and tensile yield strength values.

The Vickers hardness test is used for certain purposes, but for routine testing has been generally superseded by the Rockwell method. The former is claimed by some workers to be a more sensitive test and is favoured for investigational and fundamental studies. The Vickers test is also based on indentation under controlled loading conditions, the hardness in this case being indicated by the diagonal length of the indentation rather than its depth. It employs a square-based pyramid diamond indenter having an included angle of 136° between opposite faces. The loading system is based on weights with levers and cams and the load can range from 1 kgf up to 100 kgf.

Bend tests. A number of bend tests have also been used in the past to measure bending behaviour. The most popular in the United Kingdom was the Jenkins alternating bend test in which the number of reverse bends a ½ in wide specimen could withstand when bent through 180° around a mandrel up to the point of fracture was determined. This test had been in use for a number of years on pack rolled plate, and was useful for that material: in particular, it gave a useful measure of directionality, which was quoted as the ratio of bend values parallel to and at right angles to the direction of rolling. As this property became less significant, with the advent of cold reduced plate, the test became obsolete. Several other bend tests have also been used in the past. Descriptions of these obsolescent tests are given in the literature.[1]

Cupping tests. Several types of cupping tests aiming to provide a measure of ductility have also been used; most of these were based on the principle of deforming a specimen clamped between two circular faces by means of a hemispherically nosed punch, propelled into the specimen mechanically or hydraulically to the point of fracture. In Europe the Erichsen cupping test was the most popular of this type, whereas in the United States the Olsen test was the more favoured. In reality, these tests simply measured the stretching behaviour of the plate. Many attempts were made to refine this type of test by controlling blank-holding pressure, adopting different punch forms, etc. Many deep drawing tests have also been developed which simulate fairly closely the deformation conditions obtaining in a deep drawing press, e.g. the Swift test and its many modifications. Press-forming operations vary widely in the mode of metal deformation, and there is no single test which will predict with any degree of certainty likely behaviour in all types of press work. A useful classification of types of drawing operation has been proposed by several workers, including Wilson,[(2)] who suggested three categories: (i) pure stretch forming, (ii) pure deep drawing and (iii) a combination of these two basic forms. A large proportion of practical press work will consist of both stretching and deep-drawing, and will therefore belong to category (iii). Many attempts have also been made to correlate tensile derived properties with drawing behaviour, with various degrees of success. Press operations are discussed in detail in the chapter on canmaking.

Nowadays, reliance is largely placed on the plate manufacture's ability to maintain the well-proven practice of using the appropriate grade of steel, essentially free of casting defects, and to control adequately all processing conditions (hot and cold rolling, annealing, and temper rolling) to provide plate of acceptable drawing behaviour.

With the development of the newer drawing processes in the canmaking field, however, more and more use is made of the "special" tensile test parameters, normal anisotropy (r) and work hardening coefficient (n) together with earing propensity and ferrite grain size; in these cases, particular co-operation is called for between manufacturer and canmaker. Application of these test features is also discussed in the chapter on canmaking.

3.3.2 Double reduced plate

Again there is no single test or group of tests which give a wholly reliable prediction of the fabricating peformance of double reduced plate. The primary consideration is that the plate performs satisfactorily in the particular canmaking operation.

The most valuable results are obtained from the conventional tensile test, but this requires great care and skill, particularly for the relatively thin gauges involved; it is also time-consuming. Two other tests are used to give an approximate indication of properties. These alternatives are more rapidly carried out, but should only be used in agreement with the manufacturer. One is the Springback Tester, developed by the Continental Can Corporation of America for this specific purpose, and it gives a fairly accurate value for proof stress. The basis of the test and its method of operation are described by O'Donnell.(3) The other is the Rockwell Superficial Hardness Tester, as used for single reduced plate.

If a dispute should occur between the manufacturer and the purchaser in respect of the true test value, only the tensile test is acceptable as the referee method.

The three strength grades of double reduced plate commercially available at present are given in Table 3.3.3.

TABLE 3.3.3. TINPLATE TEMPER DESIGNATIONS: DOUBLE REDUCED

Temper classification		HR30T Hardness aim				
Current	Former	Mean	Max. deviation of sample av.	Approx. UTS N/mm^2	Formability	Typical usage
DR550	DR8	73*	+3 −3	550±70†	Double reduced	Round can bodies and can ends.
DR620	DR9	76*	+3 −3	620±70†	Double reduced	Round can bodies and can ends.
DR660	DR9M	77*	+3 −3	660±70†	Double reduced	Beer and carbonated beverage can ends.

* The UTS values for single reduced tinplate (Table 3.3.1) and the HR30T values for double reduced tinplate are given for guidance only. These values do not appear in any specifications.

† The tensile values for double reduced tinplate are the aim proof stress (0.2% non-proportional elongation).

As mentioned in Section 2.11 these high strengths have been achieved by heavy cold reduction after the final annealing operation; this results in an appreciable loss of ductility. To avoid certain types of plate failure, therefore, particularly that of split flanges when the rims of a cylindrical can body are flanged over, it is recommended that the rolling direction (the "grain" direction) be around the circumference of the can. This is positively ensured by specifying the rolling direction in relation to ordered sheet size (because it invariably is rectangular and is cut up into blanks for individual can bodies "one" direction).

(1) *Tensile testing*

The tensile test is carried out in accordance with International Standard 86 (tensile testing of sheet and strip less than 3 mm and not less than 0.5 mm thick), or the Euronorm 11 (tensile test on steel sheets and strips less than 3 mm thick), in either case with additional precautions. These are given below, and are especially important because of the lower thicknesses commonly employed for double reduced plate:

(i) Specimen edges must be entirely free of discernible burrs; this can be ensured by finishing the preparation of the edges carefully with fine emery paper.
(ii) The gauge length (which is the effective length with parallel edges) must be within the tolerance 50 mm ± 0.5 mm, and the width within 12.5 mm ± 1 mm.
(iii) The rate of straining will be 1 mm/min.
(iv) The grips employed must be capable of securing the specimen in such a manner that no skew occurs during straining, to ensure that the applied stress is aligned centrally along the major axis of the test specimen.

The value recorded is the proof stress Rp 0.2 (i.e. the 0.2% non-proportional elongation); the term tensile yield strength (0.2% offset) normally used in North America is identical.

Two test specimens are taken, from the positions indicated in Fig. 3.1.1, from each of the two sample sheets taken from each bulk package sampled. The arithmetic mean of all test results recorded is to be taken as the proof strength of that batch of plate; individual test results have no meaning in relation to any specification.

(2) *Springback test* (Fig. 3.3.3.)

This test is carried out on the Springback Temper tester, model G67; the procedure for testing is described in detail in a pamphlet supplied with the machine, and also in the paper by O'Donnell referred to above. It is essential that the operational instructions be observed very carefully. Basically, the test consists in forming the specimen (6 × 1 in) 180° around a 25 mm mandrel by means of the pivoted hand-roller, and allowing it to spring back; to minimize local deformation, contact between the roller and specimen is made through a flexible shim. The radius of curvature after springback is defined as:

$$\frac{\theta}{180} = 1 - \frac{r_o}{r_1}$$

r_o and r_1 are the radii of curvature after forming and after springback. The yield stress relates to the springback angle according to:

$$\frac{\theta}{180} = 3\left[\frac{(YS)r_o}{Et}\right] - 4\left[\frac{(YS)r_o}{Et}\right]^3 \Big/ \theta$$

YS = yield stress in lb in^{-2}, E = elastic modules in lb in^{-2}, θ = springback angle in °, t = plate thickness in in.

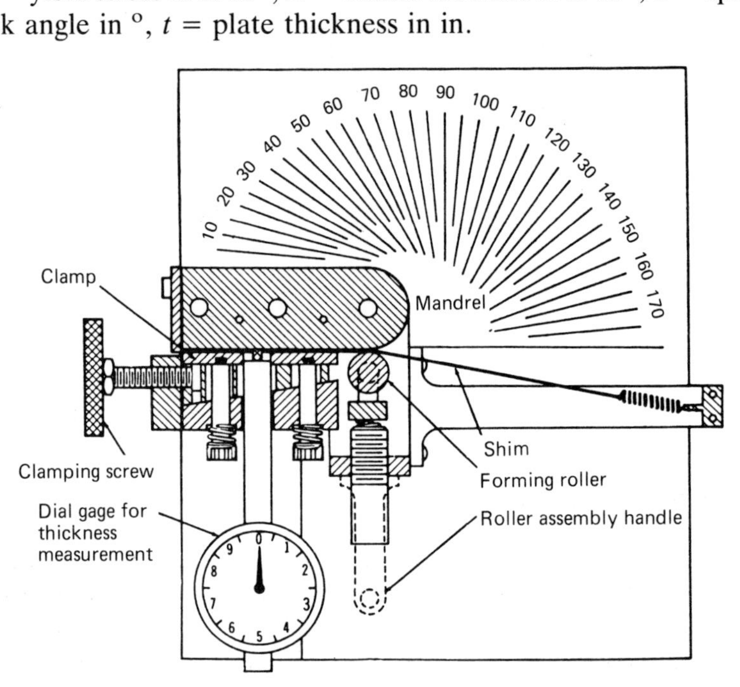

FIGURE 3.3.3 SPRING-BACK BEND TEST MACHINE

It is advisable that the conversion formula used should be agreed between manufacturer and purchaser. The number of test specimens to be taken from all sample sheets selected are identical to those described under tensile testing.

It is not necessary to remove the tin or other metallic coating prior to tensile testing or springback testing, but any organic coating — such as lacquers, enamels, printing inks or varnishes — should first be removed. It should be noted that the stoving operation used to cure these coatings after application may affect the mechanical properties of the plate; this is particularly so when the steel sheet has not fully age-hardened. Blackplate and "tin-free" steel are particular examples, as these products do not undergo a light heat-treatment after plating, as does flow-brightened tinplate, which involves heating to a temperature above the melting point of tin for a short time.

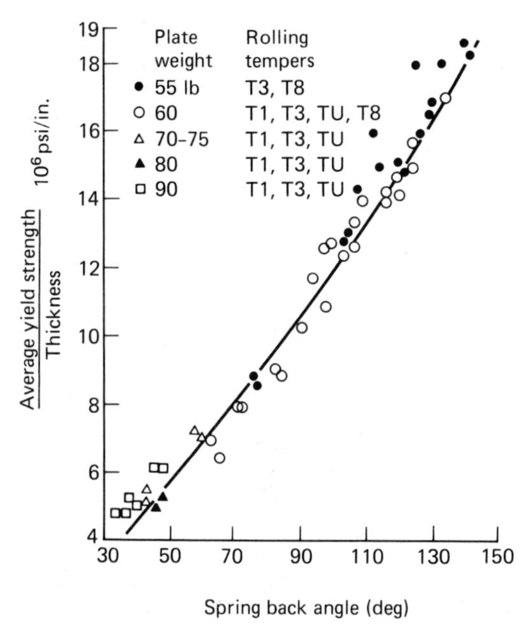

FIGURE 3.3.4 RELATIONSHIP OF SPRING-BACK ANGLE WITH YIELD STRENGTH/ THICKNESS RATIO

(3) *Hardness testing*

This hardness test is carried out exactly as that applied to single reduced plate. As a high proportion of the double reduced plate used commercially has a gauge of less than 0.22 mm, the Rockwell 15T scale (major load 15 kgf) is generally used in this case, the result being converted to HR30T values by means of the conversion table 3.3.3.

It will be recalled from the previous section that the test specimen must be detinned e.g. by immersion in Clarke's detinning solution until all action has ceased, followed by thorough washing and drying, before the hardness test is carried out. Electrolytic detinning methods are equally suitable. Any other metallic coating on the surface must be similarly removed prior to testing.

The DR 550 grade — formerly termed DR 8 — is frequently used for the shorter round built-up bodies and also for the ends of cans not subjected to high internal pressure, e.g. "dry" products and processed foods. Typical exceptions are the steel ends of beverage cans for both beer and soft drinks, where the stronger grades DR 620 or DR 660 (DR 9 or DR 9M) are employed to provide as much resistance as possible to the high pressures obtaining during pasteurisation of the contents and during high temperature storage.

At the time of writing, double reduced plate is not used to any extent for the manufacture of drawn components; tests are continually being made to widen the application of this type of plate. These often include changes to the design of components and also to the manufacturing process (tool design, lubrication systems, etc.).

3.4 TINCOATING PROPERTIES

Electrolytic tinplating (section 2.12.) has been developed over the last 50 years into a very high-speed highly automated process, capable of precise control of all plating conditions. The various grades of hot-dipped tinplate previously made in the United Kingdom can be obtained, as a matter of interest, from the British Standard BS 2920 : 1973.

The tin used for plating in either process is not less than 99.75 pure, and in the United Kingdom it conforms to Grade T2 as defined in BS 3253; in addition to specifying the minimum tin content, this also lays down maxima for a number of metallic constituents individually and jointly, particularly lead, antimony, bismuth, copper and iron. A slightly different composition may be used in some countries, but its purity should be not less than 99.75% min.

Tables 3.4.1 and 3.4.2 indicate the standard grades of electrolytic tinplate available in most countries, together with their nominal tincoating masses and the tolerances laid down. The amount of tin applied has been expressed in various ways over the years, but the term mass has now come into general use, in line with the widespread acceptance of SI units. As shown in the table, it is quoted in terms of grammes per square metre. The range of masses available is irrespective of which of the three plating electrolyte types is used by the manufacturer. Most of these masses are appreciably lower than the hot-dipped grades formerly available.

This range covers most user requirements with respect to atmospheric corrosion and product compatibility, often in combination with lacquering of the plate surface to improve its performance further. In addition to corrosion performance, soldering behaviour is also improved with increasing tincoating mass. The heaviest coated grades are generally used for packing the more corrosive fruits and vegetables, and for use when filled containers are to be stored under severe atmospheric conditions (high humidities with fluctuating temperature). The latter will usually be combined with external lacquering of the container. In view of the high cost of tin the more lightly coated grades are appreciably cheaper and are adequate for packing a number of relatively inert "dry" products and vegetables, especially when the containers are to be stored in temperate regions. A fuller discussion is given in Chapter 7, Corrosion.

TABLE 3.4.1. TINCOATING MASSES : EQUALLY COATED

Code[1]	Nominal coating mass g/m²		Minimum average coating mass[2] g/m²
	per surface	total both surfaces	total both surfaces
E2.8/2.8	2.8	5.6	4.9
E5.6/5.6	5.6	11.2	10.5
E8.4/8.4	8.4	16.8	15.7
E11.2/11.2	11.2	22.4	20.2

TABLE 3.4.2. TINCOATING MASSES : DIFFERENTIALLY COATED

Code[1]	Nominal coating mass g/m²		Minimum average[2] coating mass g/m²	
	Heavily coated surface	Lightly coated surface	Heavily coated surface	Lightly coated surface
D5.6/2.8	5.6	2.8	4.75	2.25
D8.4/2.8	8.4	2.8	7.85	2.25
D8.4/5.6	8.4	5.6	7.85	4.75
D11.2/2.8	11.2	2.8	10.1	2.25
D11.2/5.6	11.2	5.6	10.1	4.75

[1] The code figures are derived from the coating mass on each surface.
[2] Minimum average coating masses apply only to the sample taken as laid down in the specifications.

As tin is relatively expensive, attempts are continually being made to test the possibilities of reducing the mass of tin used for particular applications (for example, E8.4/8.4 in place of E11.2/11.2 for some corrosive fruits). Indeed these aims have been extended in an attempt to reduce the nominal weight specified for the quoted grades; the first example is to adopt 5.0g/m^2 in place of 5.6 g/m^2 for the grade E2.8/2.8. Many reductions have been made possible by improvements to canmaking methods, product formulation, and storage conditions.

As with the testing of all plate characteristics, it is essential that the sample collected to represent a batch of plate is taken very carefully as laid down, so that it is truly representative (see section 3.1).

It must be emphasised that the arithmetic mean of all specimens tested must be considered; individual specimen values have no relevance in relation to a specification tolerance.

The mass of tincoating applied may be determined by any of a number of methods, chemical or electrochemical, recognised in the trade. In the event of disagreement between the manufacturer and purchaser concerning the mass present on the plate, however, the only referee method accepted is the standard iodine titration procedure, which is described in detail in all current standards, e.g. Euronorm 146-180 and in the ITRI's *A Guide to Tinplate*[4]. Additionally, it is essential to take two further samples from the batch when a retest is necessary. Differential electrolytic tinplate has to be treated slightly differently during analysis, to provide the mass of tin on each surface separately, a "resist" (an air-drying varnish) is applied to one surface for the first determination, and is then removed with a suitable solvent for the second.

The complete coating will consist of the following layers:

a tin/iron alloy layer, principally $FeSn_2$ adjacent to the steel base;
an alloyed tin;
a film of mixed oxides, formed by the passivation process;
an oil film.

The compound layer, described as $FeSn_2$, is variable in crystallite size, shape, and orientation, and in its continuity. The "free" tincoating contains a large number of pores, the density per unit area being roughly inversely related to the mass. Thus the coating on tinplate presents a particularly complex system which affects many behavioural aspects; these include various forms of corrosion, solderability, lacquering and printing quality, and feeding mobility. Numerous and varied tests have been developed over the years for examination of fundamental characteristics; those currently used are based on modern physical and chemical principles, and are referred to in section 3.7. Some are described in detail in the relevant chapters on canmaking techniques, corrosion or on lacquering and printing.

3.5 ALTERNATIVE METALLIC COATINGS

The only alternative metallic coating to tin applied commercially at present in this field of materials is that based on mixtures of chromium and chromium oxides. The specified masses of chromium metal and chromium oxides vary considerably; they range broadly from 50 to 150 mg/m^2 of metallic chromium and 5 to 80 mg/m^2 of chromium oxides. These coatings are usually applied on modified electrolytic tinning lines as the process is somewhat similar, but using a different electrolyte (chromic acid). Some are "one-stage", whilst a number are "two-stage" (deposition of chromium metal followed by chromium oxides). There is no comparable coating melting stage to "flow brightening"; and the appearance of the coating is different from that of tinplate, being less bright and having a faint blue lustre. It is available on single and double reduced plate and in several surface finishes (bright, stone, and matte). Many different coating masses are offered commercially throughout the world; a short standardised list of grades is being debated, but this is not available at the time of writing.

Numerous methods of analysis for coating masses are also in use, and attempts are being made by various authorities to reach agreement on referee procedures. The two constituents of the coating are determined separately, the units used being mg/m^2. Generally the layer of chromium oxides which lies over the metallic chromium is stripped off in a solution in which the metal is virtually insoluble, and the amount in solution is determined. This is followed by determination of chromium metal coulometrically (electrolytic stripping).

3.6 LINEAR DIMENSIONS, GAUGES AND OTHER GEOMETRIC FEATURES

3.6.1 Introduction

Tinplate and other types of plate can be supplied in sheet form or in coils for subsequent cutting up into sheet by the user or direct press feeding. The majority of coils cut up by the user are sheared into the form known as "scrolled" sheet, which results in an appreciable saving in material cost, especially when large round or oval components are to be produced. The scrolled edge of the sheet is contoured closely to the shape of the blank required, thereby reducing substantially the area of scrap trim in comparison with that from a rectangular sheet.

When tinplate is supplied in sheet form, it is generally rectangular. National and International standards give specifications for rectangular sheets and coils only; scrolled sheets are not covered, as these usually fall within the users' province.

Tolerances on the physical dimensions of the plate must be closely adhered to if the highest canmaking efficiencies and accuracies which modern canmaking equipment is capable of are to be achieved. The detailed tests usually applied to ensure that these requirements are met are described in the ensuing sections.

3.6.2 Sheet dimensions and shape

Sheet sizes commonly available range up to 1000 mm in each direction; the maximum will vary according to the specification ordered, e.g. gauge, method of annealing, whether single or double reduced, and of course the type of rolling and other equipment employed by the manufacturer.

Clearly each sheet must be of adequate size to allow a true rectangle of the dimensions specified to be cut from it by the canmaker; it is common practice to shear and side trim sheets from the strip at the tinplate plant slightly larger than the ordered dimensions. Normal limits on both dimensions are up to 3 mm greater than those ordered, with an absolute maximum of 5 mm. Sheet size is checked by laying it on a flat surface and measuring the length and width along each middle line, to the nearest 0.5 mm. The sheared length, i.e. that parallel to the rolling direction, can be longer or shorter than the edge trimmed width, and conventional terminology for length and width is not uniform. The longer dimension of the sheet is more usually termed its length, even though that may have been the width of the strip prior to shearing. But it is sometimes the practice to refer to the sheared length as the sheet length (and therefore the other dimension as the sheet width, that which was the strip width); under this convention the length could be shorter than the width! This can be particularly misleading when the direction of the "grain" needs to be specified, as is often the case with double reduced plate. Ambiguity is avoided by specifying which of the two sheet dimensions is to be parallel to the rolling direction. In the absence of a specific customer request the manufacturer will roll the strip width as either the longer or the shorter sheet dimension, according to which is the more economic in relation to the mill and roll conditions.

In the case of delivered coil, the only dimension for which the manufacturer is responsible is the strip width; the purchaser will check this dimension before or after cutting up into sheet, according to which is the more convenient. For an adequate check on all dimensions the number of sheets taken for testing from each of the bulk packages selected should be not less than 0.5% of the total.

Two other important parameters are: "out-of-squareness" and "camber". Each of these features is illustrated in Fig. 3.6.1. Out-of-squareness is defined as the deviation of an edge from a straight line drawn at right angles to the other sheet edge. Most specifications quote for this feature a maximum of 0.15%, with an absolute maximum of 0.25%.

The definition of camber in this trade, which is probably peculiar to the tinplate industry, is the deviation of an edge from a straight line forming a chord to it, as in the formula

$$\frac{\text{deviation } (D)}{\text{length of chord } (L)} \times 100\%$$

Maximum camber must not exceed 0.15%.

Excessive variation of each of these two parameters can seriously affect canmaking operations and accuracy of components.

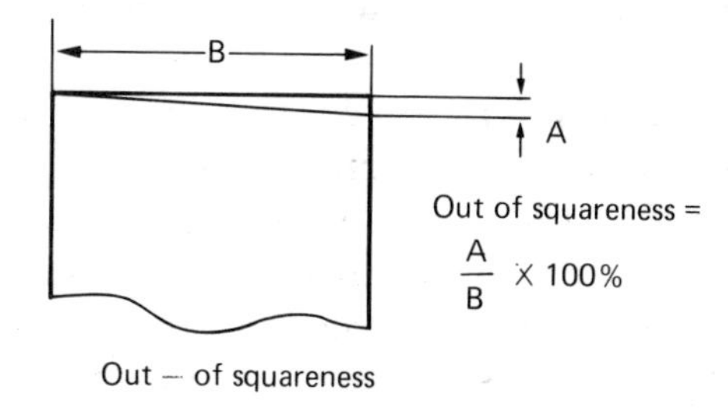

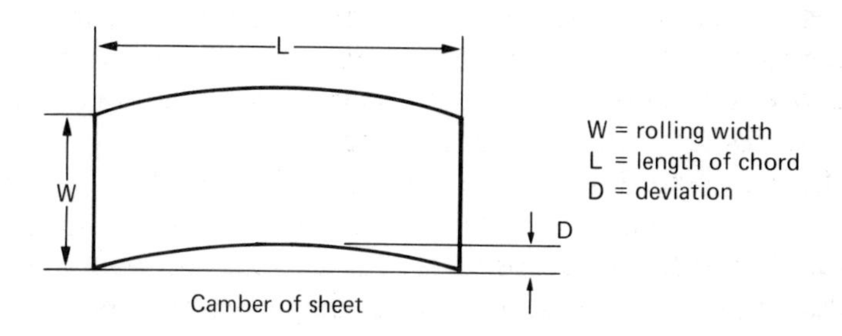

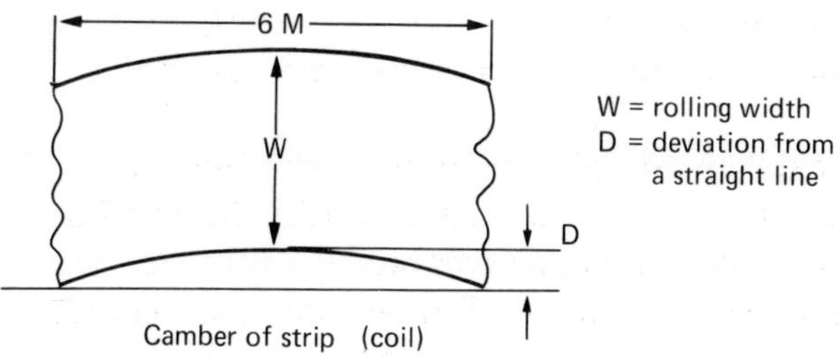

FIGURE 3.6.1 DIAGRAMMATIC REPRESENTATION OF OUT-OF-SQUARENESS AND CAMBER

In the case of sheets which have been cut from coil by the purchaser, the samples must be taken from bulk packages containing standard grade only. Width is again measured to the nearest 0.5 mm and a tolerance of not more than 3 mm greater than the ordered width is specified. Camber is based on a strip length of 6 m, and must not exceed 0.1%, i.e. 6 mm. A further feature specified in the case of coil is lateral weave, sometimes referred to as short pitch camber; this is the deviation of the mill trimmed edge from a straight line lying in the same plane and forming a chord to it over a relatively short distance. When measured over a chord length of 1 m, the lateral weave must not exceed 1.5 mm (some call for a maximum of 1.0 mm). If the coil is to be scroll sheared, the stipulated maximum tolerance may be even tighter, by agreement with the manufacturer.

There are in fact four further geometric features which can cause canmaking problems; these are given below.

edge wave: an intermittent vertical displacement occurring at the sheet (or strip) edge, when the sheet or strip is laid on a flat surface.

centre buckle (full centre): an intermittent vertical displacement or wave in the sheet or strip occurring other than at the edges.

longitudinal bow (line bow): a residual curvature in the sheet in the direction of rolling (or after the strip has been uncoiled).

transverse bow (cross bow): a mode of curvature in the sheet or strip such that the distance between the edges parallel to the rolling direction is less than the sheet width (or the strip width).

At the time of writing it is not possible to put forward precise methods of measuring these geometric features, and therefore tolerances cannot yet be specified. Manufacturers always endeavour to keep the incidence and the extent of these features to the low level needed by canmakers.

3.6.3 Gauges and edge profiles

The gauges normally available commercially are:

Single reduced plate   0.15 mm to 0.49 mm

Double reduced plate 0.14 mm to 0.29 mm (this range will be particularly dependent on the mechanical properties grade required)

In each case in multiples of 0.01 mm.

Uniformity of thickness is important, not only because of its effects on efficiency of canmaking operations, but also because of the real effects gauge variation has on several can properties, particularly bursting pressure and resistance to damage when roughly handled. As it is vital, because of the cost effect, not to over-engineer the container, detailed thickness tolerances have been agreed to ensure satisfactory can manufacture and performance. As the thickness levels generally employed are relatively low, it is more accurate to determine actual values by weighing a specimen of known area and, by means of the following formula, to convert weight into thickness:

$$\text{thickness (mm)} = \frac{\text{mass (g)}}{\text{actual area (mm}^2) \times 0.00785}$$

Direct measurement of thickness, provided a particular type of micrometer is used, in acceptable between manufacturer and purchaser in some countries. That usually recommended is hand operated and spring loaded, and must be accurate to 0.001 mm. It has a ball-ended shank anvil about 3 mm diameter, a curved surface base anvil of approximately 25 mm radius and a face approximately 13 mm diameter.

When the average thickness of a whole sheet is required, it is weighed to a precision of 2 g and its dimensions are measured to the nearest 0.5 mm. In a similar manner, variation in thickness within a sheet is determined by weighing the two specimens coded Y in Fig. 3.1.1. These specimens are weighed to the nearest 0.01 g and their dimensions measured to the nearest 0.1 mm. In each case, average thickness is calculated by means of the above formula. As plate gauge is available in steps of 0.01 mm, and variation in thickness overall is of the order of 10%, it is the general practice to quote thickness determinations to the nearest 0.001 mm.

The tolerances on thickness variation agreed in European and ISO standards are as follows:

In sheet form:

(i) consignment means –
more than 20 000 sheets ± 2.5% of nominal thickness
20 000 sheets or less ± 4% of nominal thickness

(ii) individual sheets – none outside the limits ± 8.5%

(iii) within sheets – the thickness of either of the two specimens Y in Fig. 3.1.1. must not deviate from the actual average thickness of the whole sheet by more than ± 4%.

Conditions (i) to (iii) above apply to material supplied in coiled form also; the user will extract sample sheets in a similar manner after cutting up the coil, and then determine gauge as done on plate supplied in sheet form. In both cases, conditions (ii) and (iii) refer to individual sheets, provided they are taken as a random sample in the agreed manner. Condition (i), however, refers essentially to the arithmetic mean of all sheets tested, after selecting according to the approved method of sampling. Coiled material, which will be fed directly to a press, presents a special case, and sampling procedures to be adopted in such cases should be agreed by both purchaser and manufacturer.

In the case of double reduced plate, a further stipulation is imposed, to limit the extent of thinning adjacent to the trimmed rolling edge, in comparison with the thickness at the centre of the sheet. This thinning is referred to as the "feather edge" or the transverse thickness profile. It is, of course, possible for the edge to be thicker than the centre, but this is rare; as can be expected from the rolling conditions, the normal profile at right angles to the rolling direction is for a slightly thinner edge than the centre of the strip. As "edge" thickness is measured, by agreement, at a point 6 mm inwards from the trimmed edge, the particular type of micrometer described above must be used at these two transverse locations. Some standards specify for this variation a minimum thickness at the edge location as defined as minus 8% of the centre thickness of the particular sheet being measured. Some national standards, however, quote a minimum thickness at that point of not more than 15% below the specified nominal thickness. This feature is important with single reduced plate, but is especially so in the case of double reduced plate, as the thinner edges of this less ductile material can split and dent more readily when sheets or slit blanks are being mechanically fed through high-speed canmaking machinery. This mode of failure is also dependent on the quality of the slit edge; a rough or badly burred edge appreciably increases the likelihood of splitting during feeding or stretching in a flanging operation.

Burr can be described as metal displaced beyond the plane of the sheet surface by the shearing action of slitting; its extent can vary significantly according to the sharpness and setting of the rotary knives which are used. Some standards specify a maximum of 0.05

mm (0.002 in), but manufacturers should endeavour always to maintain the burr at a minimum, particularly in the case of double reduced plate.

3.7 OTHER TESTING PROCEDURES

In addition to the test procedures described earlier, which must be applied to check conformity of a batch of plate with a specification, there are numerous other tests available for investigational or fundamental research purposes. Most of these are complex and demand considerable expertise, in addition to needing sophisticated apparatus. None of these as yet is included in any standard. They range over a wide field:

(i) metallographic and physical testing of the steel base;
(ii) microscopic and electrochemical testing of the tin-iron alloy layer existing between the steel base surface and the tin-coating;
(iii) metallographic and electrochemical studies of the tincoating layer;
(iv) comprehensive studies of the physical and chemical characteristics of the surface, including the so-called passivation layer, and its modification by the atmospheric conditions obtaining during storage, and of the applied oil film.

As would be expected, many of the testing procedures which have been developed over the years for particular purposes are no longer used, but are of some academic interest. Others have proved of immense value in developing knowledge of the complex structure and properties of tinplate and its variants, and their relationship with performance. The development of the electron microscope and its modifications, modern electrochemical techniques, and advanced tests for study of metal deformation have been especially helpful.

Each layer in the structure of tinplate has an important relationship with more than one facet of behaviour; e.g. the composition and structure of the baseplate affect its deformation characteristics, its strength and corrosion behaviour in various media and under atmospheric conditions. Detailed description of the large number of tests employed in these areas is outside the scope of this book, but the more "significant" ones are described in later chapters; e.g. tests of solderability behaviour, and metal deformation under stress coupled with the related plate characteristics in Chapter 4 on canmaking processes; tests to assess the likely corrosion behaviour in given environments in Chapter 7; and the influence of surface factors on lacquerability and printing performance in Chapter 6.

The reader is directed to literature references for further information: plate manufacturers and major canmakers, together with the International Tin Research Institute,[4] have extensive libraries and information services. A general account of specialised tests is given in Appendix II.

3.8 STANDARDISATION AND PUBLISHED STANDARDS

Considerable effort has been made over a number of years towards the improvement and enlargement of standard specifications for these materials, and these efforts are likely to continue on a wide front for sometime. The first British Standard describing tinplate and blackplate in sheet form (BS 2920) was published in 1957, and has since been improved and updated; the current standard is BS 2920: 1973. Efforts to standardise tinplate-type materials internationally were set in motion by the International Standards Organisation (based in Geneva) in 1960, while similar activities were promoted in the European field by the European Coal and Steel Community in the early 1950s. Considerable impetus was given to the latter's activities in 1973 when the United Kingdom, together with Denmark and Ireland, joined the original six member states; this enlargement of the EEC resulted in a new technical committee being formed. The British Standards Institute, who already at that time held the Secretariat of the relevant ISO committee, were invited to serve as the Secretariat of the new ECSC technical committee. Several standard specifications have already been published by both organisations (details are given later); further documents to cover other facets of plate material and quality are in the course of preparation, and will be published in the near future.

TABLE 3.8.1. TINPLATE STANDARDS

National Standards			
Country	**No.**	**Year**	**Title**
Australia	AS 1517 Pt.1 AS 1517 Pt.2	1982 1982	Tinplate and blackplate Pt.1 sheet Tinplate and blackplate Pt.2 coil
Austria	M 3422	1964	Pack rolled tinplate - dimensions and tolerances
Belgium	NBN 668	1966	Blackplate and tinplate - quality requirements

Country	Standard	Year	Title
	NBN 669	1966	Blackplate and tinplate tolerances
Colombia	1 CONTEC 647	1978	Tinplate
France	NF A 36-150	1970	Tinplate and blackplate - qualities
Germany F.R.	DIN 1540	1970	Blackplate and tinplate in sheets: permissible variations in dimensions and form
	DIN 1616	1981	Tinplate and plackplate in sheet form: qualities, dimensions and tolerances (supersedes DIN 1540:04)
Hungary	MSZ 4461	1972	Cold reduced tinplate
India	IS 597	1962	Blackplate for tinning and tinplate (pack rolled)
	IS 1993	1974	Specification for cold-reduced tinplate and cold reduced black-plate (1st revision)
	IS 9025	1978	Specification for cold-reduced electolytic tinplate and cold-reduced blackplate in coil form
Iran	ISRI 1291	1975	Cold-reduced tinplate and cold-reduced blackplate Pt.1-sheet
Italy	UNI 7644 (Experimental)	1976	Cold-reduced and coated flat finished steel products - Tinplate in coils for subsequent cutting into sheet form
	UNI 7645	1976	Cold-reduced and coated flat finished steel products. Tinplate in sheet form (supersedes UNI 5755)
Japan	JIS G 3303	1975	Tinplate and blackplate
	JIS H 0402	1975	Methods of test for tin plating
South Korea	KS D 3516	1974	Tinplate
Mexico	B 34	1961	Blackplate, tinplate and lead plate (terneplate) for the manufacture of containers
New Zealand		1970	Cold-reduced tinplate and cold-reduced blackplate (B.S. 2920-69 adopted by N.Z.)
Poland	PN-H 92122	1973	Tinplate
Spain	UNE 7-301	1973	Volumetric method to determine the weight of the tin coating on tinplate: titration with iodine (EU)
	UNE 36-091 Pt.1	1974	Cold-reduced tinplate and cold-reduced blackplate in sheet
	UNE 36-091 Pt.2	1978	Tinplate and blackplate in coils: tolerances and conditions for delivery
Taiwan	CNS G 3009	1971	Tinplate (hot-dip method)
	CN G 3097	1977	Electrolytic tinplate
Thailand	TIS 16	1971	Specification for cold-reduced tinplate and cold-reduced black-plate
United Kingdom	BS 2920	1973	Cold-reduced tinplate and cold-reduced blackplate
U.S.S.R.	GOST 13345	1978	Electro-coated tinplate: technical conditions

	GOST 15580	1970	Cold-reduced blackplate for hot tinning (supersedes GOST 9488 and 7530
	GOST 17718	1972	Hot-rolled, hot dipped tinplate (partially supersedes GOST 5343-54
U.S.A.	ANSI/ATSM 623-77	1977	Specification for general requirements for tin mill products
	ANSI/ASTM 623M-78	1978	Standard metric products
	ANSI/ATSM 624-78	1978	Standard specification for single-reduced electrolytic tinplate
	ANSI/ASTM A 625-78	1978	Standard specification for single-reduced blackplate
	ANSI/ATSM A 626-78	1978	Standard specification for double-reduced electrolytic tinplate
	ANSI/ATSM A 630-68	1968	Standard methods for determination of tin coating weights for hot-dip and electrolytic tinplate
	ANSI/ATSM A 650-71	1971	Standard specification for double-reduced blackplate
	Fed. Spec. QQ-T-425B	1976	Tinplate (electrolytic) (supersedes Fed. Spec. QQ-T-425A 1970)
International Standards			
Europe	EURONORM 145	1978	Tinplate and blackplate - qualities, dimensions and tolerances
	EURONORM 146	1980	Tinplate and blackplate in coil form for subsequent cutting into sheets - qualities, dimensions and tolerances
International	ISO R1111/1	1969	Cold-reduced tinplate and cold-reduced blackplate Pt.1-sheet
	ISO R1111/2	1976	Cold-reduced tinplate and cold-reduced blackplate Pt.2. Coil for subsequent cutting into sheet form
Pan American	COPANT 691	1976	Tinplate

TABLE 3.8.2. RELATED STANDARDS

European:	Euronorm 109-72	Conventional Rockwell Hardness tests. Rockwell scales HRN and HRT. Rockwell scales HRBm and HRTm for thin products
	Euronorm 11	Tensile test on steel sheets and strips less than 3 mm thick
International Standards:	ISO 3798-76	Tinplate and Blackplate - Minimum Packaging Requirements
	ISO 1024-69	Rockwell Superficial Hardness Test for Steel (N and T scales)
	ISO 86-74	Tensile testing of sheet and strip less than 3 mm thick

Standards likely to be published over the next few years:	
	Double reduced tinplate and blackplate in sheet and in coil form
As ISO and as Euronorm Standards	Chromium/chromium oxide coated steel, sheet in single reduced and double reduced varieties, in sheet and in coil forms

Copies of any Standard may be obtained in the UK from the British Standards Institution.

These committees were established with the particular aim of including knowledgeable technical representatives of all major manufacturing and user groups, together with experts from independent research organisations and from as many member states as possible. Great care is taken to include consideration of all interests and points of view, in addition to basic technical criteria.

There are many and varied reasons for the need to develop better harmonised and more comprehensive international standards. They are valuable aids towards removal of technical barriers to trade, which could otherwise arise from differences in practices and definitions between nations. They help to promote better understanding between manufacturers and with users. They are of greatest advantage to the smaller and less experienced users. It is also, of course, a particular object of international discussion to eliminate differences between international and national standards. The exchange of views and the criteria contained in better harmonised standards are of undoubted help towards more precise definition of customer requirements, as well as better understanding of manufacturers' capabilities.

New and improved canmaking techniques usually require material having modified characteristics, the specification of which is likely ultimately to improve understanding of material behaviour and to more comprehensive specifications.

Aspects covered in standardisation are:

1. Definition of tinplate and other types of plate used in the container industry.
2. Material grading, according to visual appearance and defects, dimensional and thickness tolerance.
3. Tincoating masses.
4. Mechanical properties (Temper).
5. Dimensional requirements; sizes, shape, and thickness.
6. Sampling and testing procedures; tincoating masses, mechanical properties, detailed description of test methods and retesting.

7. Coding systems, packaging methods, and special arrangements between manufacturers and purchasers.
8. Glossary of terms.

APPENDIX I

CLARKE'S DETINNING METHOD

By this method, the tincoating and tin-iron alloy layer are completely removed by immersion in Clarke's detinning solution;[5] this consists of concentrated hydrochloric acid inhibited with Sb_2Cl_3. The specimen is removed from the solution about 1 minute after all gas evolution has ceased, then thoroughly washed in running water; the loosely adherent antimony is swabbed off with cotton wool, and the specimen finally dried.

The solution is prepared by adding 20g antimony trioxide (Sb_2O_3) to 1 litre of concentrated hydrochloric acid.

If the method is used as a rapid check on total tincoating weight, a correction must be applied for the very small amount of iron also dissolved; the total weight is determined by accurate weighing of a specimen of determined area before and after removing the tin.

APPENDIX II

SPECIALISED TESTING TECHNIQUES

Many types of modern sophisticated testing apparatus are available nowadays to study in depth several aspects of tinplate and container behaviour; these apply to metallic surfaces/coatings; composition of metal in bulk form; metallic contents in liquid foods and beverages; and identification/analysis of organic materials and gases, such as lacquers, and the headspace gases in packed cans. Brief descriptions of the scientific equipment used and the basis of their operation are given in the attached table.

Many other techniques, now firmly established, are based on coulometric determinations, in which measurement is made of the charge passed through an electrolysis cell during particular oxidation or reduction operations. Some examples of their application are: tincoating weight determination; the passivation layer, for tin oxide, chromium and chromium oxides amounts. The apparatus consists basically of a cell containing the sample to be tested, selected electrolyte, means for passing through various gases to purge and maintain anaerobic conditions, the cell being otherwise fully sealed. A power source capable of supplying controlled constant current in the ranges normally 25 μA up to 200 mA is employed, together with voltage recorders covering several potential ranges. Many forms of commercial apparatus are available for these purposes, which allow selection of current density, current direction, several voltage ranges, coupled with automatic control of gaseous purging.

MODERN SCIENTIFIC EQUIPMENT USED IN SPECIAL INVESTIGATIONAL STUDIES

Type of examination	Specialist equipment	Basis of test
Metallic Surfaces/Coatings	*Stereo-Scan* with a link System *Stereo-Scan* with crystal x-ray analyser *X-ray fluorescence spectroscope* capable of scanning 1 mm dia. fields of metallic coatings on ferrous substrate (e.g. tin coatings)	Bombardment of surfaces with high voltage electrons and measurement of secondary electron emission.
	X-ray fluoroscent spectroscope for macro analysis (15 mm dia. field) of metallic coatings on ferrous substrate - not suitable for scanning	Measurement of x-ray scatter/absorbtion by coatings. Each type of coating and coating thickness being first calibrated against x-ray emission from pure iron.
	X-ray photo-electron spectroscope (XPS) for analysis of very thin (about 30A) metallic and non-metallic surface layers *Electron microscope*	Bombardment by x-rays and measurement of energy of emitted electrons from different elements
Analysis of metals in bulk form (e.g. tinplate steels, tool steels)	Atomic emission spectroscope	Atomising of metallic and non-metallic elements by means of an electric arc, and measurement of the energy emitted by excited atoms of different elements present in the test sample.
Analysis of metals in liquids e.g. in foods/beverages in trace amounts	atomic absorbtion spectrometer	Atomising elemental metals by means of flame or electric charge and measurement of the energy absorbed by the unexcited atoms. The absorbed energy being proportional to the concentration of different metals in the test sample.
Identification/ analysis of organic materials and gases e.g. lacquers and head space gases in cans	1. Gas/liquid chromatograph 2. Infra-red spectrophotometer	1. Selective measurements of the rates of travel of different gases/liquids through a chromatographic column. 2. Exposure to infra-red energy and recording the energy absorbed by different compounds in terms of wave lengths associated with different compounds.

3. Ultraviolet spectrophotometer	3. Exposure of samples to ultraviolet energy and measurement of absorbed energy by different types of molecules.
4. Mass spectrograph	4. Bombardment of molecules with high energy electrons. Electromagnetic separation of the positively charged molecules with respect to their charge and mass and recording these using u.v. recorder chart.

REFERENCES

1. W.E. HOARE *et al, The Technology of Tinplate,* Edward Arnold, London, 1965.
2. D.W. WILSON, Development and Present Status of Formability Tests of the "Swift" Type, *Sheet Metal Ind.* **40** (432), (1963), 249
3. J.E. O'DONNEL *et al,* A New Tool for Testing Thin Sheet, *Metal Progress* **81** (5) (1962), 67.
4. *A Guide to Tinplate,* ITRI, 1983.
5. S.G. CLARKE, A Rapid Test of Thickness of Tin Coatings on Steel, *Analyst,* **59** (1934) 525.

Chapter 4

Canmaking Processes

4.1 INTRODUCTION

Extensive development of canmaking techniques and of all the raw materials used in their manufacture has taken place over the last 20 years or so; this considerable effort has been applied over a broad front, but especially so in the development of several new sophisticated drawing techniques. Most of these are now firmly established in mass production.

Although the longer established techniques — those used for making three-piece cans (which consist of a cylindrical body and two separate ends) and miscellaneous containers — have been in use since at least the beginning of this century, they have been improved in many respects. This includes manufacturing efficiency, speed and container quality, and they are still widely used. Substantial capital is involved, and as long as the processes continue to be improved to meet increasing customer demands for improved quality in terms of product integrity, coupled with resistance to abuse, customer appeal and acceptable cost, their use is likely to continue.

Quality and service demands, cost and convenience have led more particularly to many new approaches, and these have been assisted by improved quality of plate and other raw materials. Included in these are the Draw and Wall-iron (DWI) and Draw-Redraw (DRD) processes, producing two-piece bodies (a cylindrical body and bottom end integrally in one piece and not incorporating a side seam, together with a separate top end), and also sophisticated high-speed welding of the side seam in three-piece cans in place of soldering.

Before proceeding to a description of these processes, the desired properties of the containers, and the effects of plate characteristics on the process and can properties, it will be useful to give a broad account of the mechanical stresses imposed on the cans by filling, transport and storage conditions.

4.2 MECHANICAL STRENGTHS OF EMPTY AND FILLED CANS

The majority of cylindrical cans, both the two- and three-piece types, will be subjected to heavy stresses during manufacture and subsequent use, as illustrated in Fig. 4.2.1. These can be summarised as:

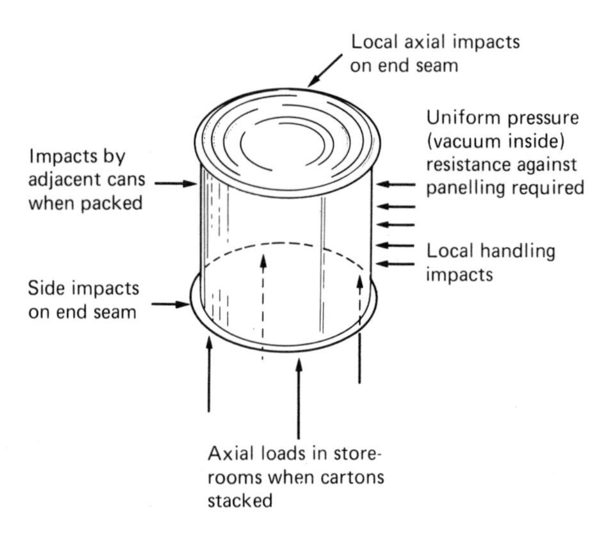

FIGURE 4.2.1 IMPACT FORCES DURING HANDLING AND TRANSPORTATION

appreciable axial loads: during the double seaming method of attaching ends to the body, and in some filling operations, during warehouse storage of empty and filled cans on pallets to a height of at least 10 m;

high internal pressures developed during sterilisation of food contents, filling and pasteurising highly carbonated beers and soft drinks, and in filling aerosol products;

some products are filled under appreciable internal vacua;

physically severe transport conditions, especially through rail journeys and general rough handling;

some products have internal pressures which are particularly temperature sensitive, and can increase substantially under conditions of high ambient temperature.

There is little doubt that some years ago cans were "over-engineered", giving an unnecessarily high margin of safety. The need for cost reduction has led to plate thickness being reduced significantly, and consequently to a reduced but adequate margin of safety.

Several methods of reducing the possibility of failure have abeen devised:

(i) Can body designs include a system of circumferential beads (Fig. 4.2.2) which will appreciably increase the resistance to denting from sideways blows or internal vacuum, but this last may be at the expense of axial collapse load. Appropriate compromise may be required by optimum bead design.

FIGURE 4.2.2 TYPICAL BEAD PROFILES

(ii) End profiles are designed to give required outward/inward deformation characteristics under internal pressures or vacua without suffering permanent deformation; superimposed external pressure can be used during filling/processing to counteract some of these dangers.

(iii) Improved packaging methods and conveyor systems are often adopted to reduce damage hazards.

Laboratory methods of evaluating these characteristics will include dynamic, hydraulic or pneumatic testing and abuse tests of individual empty or filled containers. (Pneumatic testing in particular must be carried out carefully, including adoption of safety devices, as a metal container could dismember on ultimate failure with considerable force under these conditions.) The latter would need to simulate closely commercial usage conditions; a typical procedure would consist of subjecting the relevant carton of filled cans, and also individual cans, to a series of drop tests from various heights on to hard floors, so controlling the drop that the carton falls on edges and corners, as well as on its flat faces. Assessment of failure is usually visual and subjective, but useful information is provided by this type of preliminary testing before comprehensive commercial testing jointly with users. These will include trials through customer filling and packing procedures, and finally subjecting them to transport trials by rail and road. Some authorities, such as transport and official agencies, lay down obligatory trials before new containers are accepted for conveyance.

4.3 THREE-PIECE CAN MANUFACTURE

This can consist of a cylindrical body and two ends, one of which normally will be attached to the body (the "maker's end") and the other despatched loose to the user for attachment to the other end after filling. Although some three-piece cans have a rectangular cross-section (e.g. solid meat packs), by far the majority are cylindrical. They are generally termed open-top cans in the food-packaging context; some cylindrical three-piece cans are also made for the trade known as "general line" for containing a wide range of non-food products; again, the vast majority are for processed foods. Other general line containers vary widely in design and methods of manufacture, and these will be described broadly later. The majority of open top cans vary in diameter from approximately 52 to 102 mm, together with a large can of 155 mm for "institutional" purposes. Their heights will be within a similar range, but height is usually greater than diameter.

It will be preferable to describe end manufacture first, followed by that of the body and finally attachment of end to body.

4.3.1 End manufacture

The end is a precise pressing to conform to the design, usually complex, developed for optimum deformation behaviour. A typical end design is shown in Fig. 4.3.1. In addition to plate thickness, the precise contour of the expansion rings and countersink depth have a marked effect on its deformation characteristics. The end behaves as a flexible membrane under internal and external absolute pressures, and must not deform to the extent of permanent distortion.

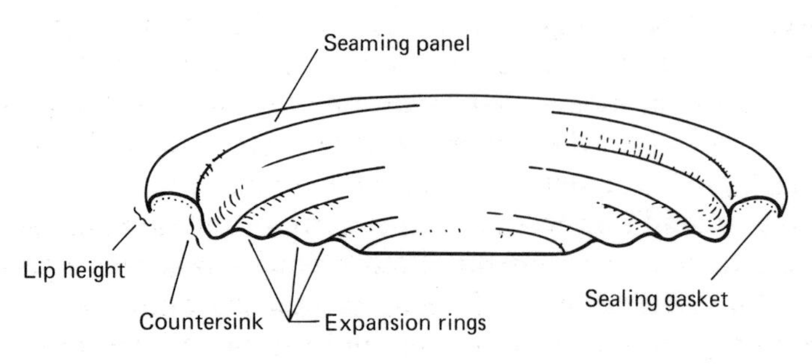

FIGURE 4.3.1 END DESIGN

Its contour is designed so that it can support the required internal pressure through tensile rather than bending stresses, and the rim is stiffened to support large circumferential compressive stresses without buckling. The end seaming panel is designed so that it can be readily rolled with the body end contour into an effective hermatic seal. Accurate tool manufacture is required to achieve the desired contour, and this also calls for precise setting in the press. Single action presses capable of giving minimum deflection under the operating loads and double die sets are usual.

The end periphery is curled inwards to assist individual separation from the stack under high production speeds, and to help in forming the double seam, the joint between end and body. It is necessary, therefore, to stamp the end without its peripheral area being contained between the punch and draw ring under controlled pressure, as shown in Fig. 4.3.2(b); (a) illustrates the optimum drawing condition. Lack of metal control in the seaming panel results in a tendency to form wrinkles or "puckers"; to avoid this fault, which would otherwise follow through into the double seam, it is usual to coin the area by heavy setting of the stamping tools. Modern presses operating at speeds up to 400 strokes per minute are normal; as mentioned earlier, a set of double dies is generally used, resulting in 800 ends per minute.

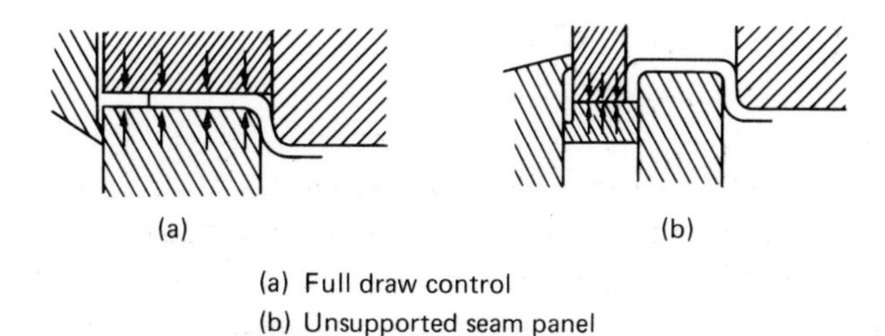

(a) Full draw control
(b) Unsupported seam panel

FIGURE 4.3.2 END STAMPING CONDITIONS

A circular blank is required for the ends. If cut from a square area of plate, considerable scrap will result. For ecomonic plate usage, the strips cut from the tinplate sheet for feeding into the press are "scrolled" in shape; the extent of saving possible is illustrated in Fig. 4.3.3(a). This layout still gives a loss of half-blank at the end of each strip, in addition to that in the narrow shreds; residual shred widths are kept to a minimum consistent with accurate high-speed feeding through the press. Increasing raw material cost demand maximum economy in plate usage, and use of coil cut-up lines for scroll shearing endplate by the canmaker is widespread. (Tinplate manufacturers do not normally offer plate in primary scrolled form.) The extra saving offered by this technique is also shown in Fig. 4.3.3(b). Substantial capital is involved, and their use demands new techniques, especially plate inspection methods (surface faults, off gauge, pinholes, etc.) and modified feeding of primary scrolled

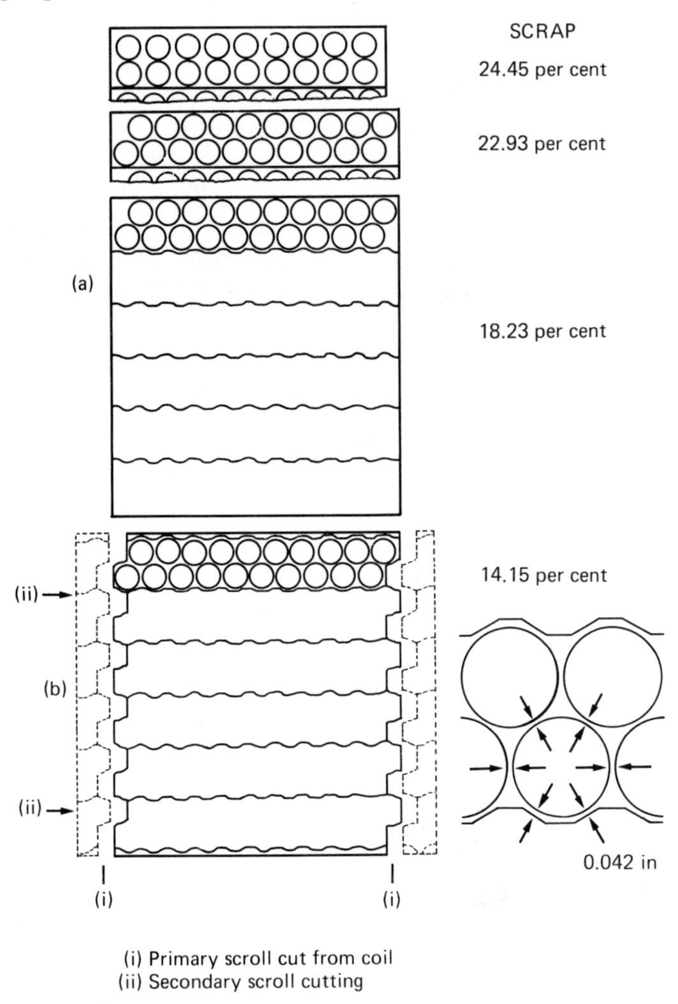

FIGURE 4.3.3 SHRED WITH ROUND STAMPINGS: (a) FROM RECTANGULAR SHEETS (b) FROM SCROLLED SHEETS

plate rather than rectangular sheets through lacquering and printing lines. (When sheet has been primarily scroll sheared, clearly secondary scroll shearing is at right angles to the primary shearing.) Some operators favour whole sheet feeding into presses, which does not involve prior strip cutting, especially for small diameter components. These are regarded as being more difficult to operate, and their adoption will depend significantly on particular conditions.

Maintaining the designed end profile on stamping requires that the plate characteristics be within tolerance in respect of gauge and temper. It was the practice some years ago, in order to maintain end profile, to reset the press when changing from batch-annealed to continuously annealed plate, in view of their generally different spring-back properties; nowadays, as higher temper grades are the norm, resetting is seldom necessary. Within gauge tolerance is vital in many respects, including curl shape, smooth feeding to ensure maximum operating efficiency, etc. Close quality control of end profile is vital.

The endplate may have been previously coated with lacquer, or used unlacquered ("plain"): in the former case a stamping lubricant is applied to minimize lacquer fracture on tooling, or alternatively an "internal" lubricant may be included in the lacquer prior to application, which often is considered adequate, particularly with good flexible lacquer types (described in Chapter 6). External lubrication is usually done by fine spraying on to the strips as they are picked up automatically with suction-cups from the hopper and fed into the press. All lubricants used must be wholly compatible with food products. Unlacquered plate is not normally lubricated on end stamping; that applied to the plate surface in the electrolytic tinning process is usually adequate.

Open frame (or "C") type presses are normally used, and although this type has a very rigid structure, the combined operation of blanking, drawing and coining sets up some frame deflection, with consequent effects on end profile. Discussion of these has been presented by Stuchbery;[1] the paper includes several graphs illustrating the deflections resulting from the operation, and also the effects of variation in plate thickness. Variation in end profile resulting from these factors are also covered. The analysis of this behaviour presented by Stuchbery suggests some improvements to profile design to decrease extent of frame deflection, coupled with reduced effects of plate thickness and tendency to wrinkling in the seaming panel. Data are presented on the effects of profile detail on the deformation characteristics of the ends — these are expressed as the distortion in its centre with varying internal absolute pressure. The critical feature, peaking pressure, is that at which the end will distort outwards sharply and permanently, usually starting at the base of the countersink; this is wholly unacceptable and every profile must be capable of withstanding the maximum differential pressure likely to be met in use without permanently distorting, i.e. it must return to its original profile without visually detectable change when the pressure is released. This failure pressure can be increased in several ways; increasing plate thickness (usually unacceptable economically), by increasing the countersink depth, and by using sharper blending radii as its base; there is a limit to which these modifications can be made as there is a danger of lacquer fracture and even of plate fracture during the stamping operation and in subsequent double seaming.

After stamping, the ends fall through the press into the curler, which consists of an inner rotating disc and an outer stationary tool, both suitably profiled, to form the outside curl and outside diameter.

A lining or sealing compound is then applied into the seaming panel through closely matched nozzles. During its rotation, the machine head picks up an end for each of its several lining stations, and discharges it to a vertical drying tower oven. Each chuck rotates at up to 3000 revs/min, and over some two revolutions, the feed nozzle releases the sealing compound into the end channel, and centrifugal force causes this to flow into the curl. The sealant used is based on natural or synthetic rubber and dispersed in water or solvent. It is formulated to possess appropriate flow characteristics which must remain stable over storage periods of many months. Its constituents are subject to stringent food regulations. Compound manufacture is a specialist operation and normally it is supplied by external organisations. For some types of end closures, e.g. screw caps for bottles, a 100% plastisol formulation may be used.

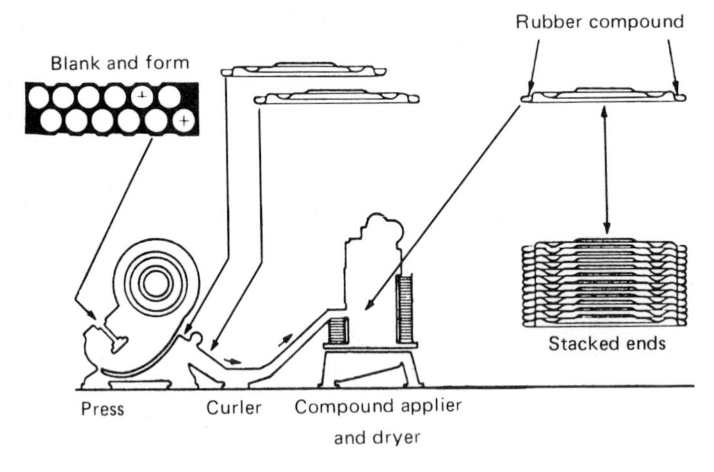

FIGURE 4.3.4 SCHEMATIC VIEW OF END STAMPING PRESS AND COMPOUND LINING MACHINE

The various units involved in end manufacture are often linked together, as illustrated diagrammatically in Fig. 4.3.4. To provide the standard required, quality checks are regularly carried out both by the machine operators and independent quality control. These will cover all dimensional features of significance and weight of sealing compound applied; go – no go gauges and other measuring instruments designed for the purpose are used. Often the quality control personnel will have the responsibility for advising stopping of production until any fault is rectified.

4.3.2 Body production

As virtually all open top can bodies require a rectangular blank for their fabrication, the waste arising from cutting up the rectangular sheet is minimal; it is confined to the narrow sheet edge trim taken off during the slitting operation. There is no real material saving, therefore, in purchasing coil, and it is the general custom to purchase body plate in sheet form. Some users' coil cut-up lines can be fitted with straight-cut knives as well as the "scrolled" type, to be in a position to cut rectangular sheets internally, should the need arise.

If the bodies are required lacquered and/or printed, these operations are carried out before cutting up the sheet.

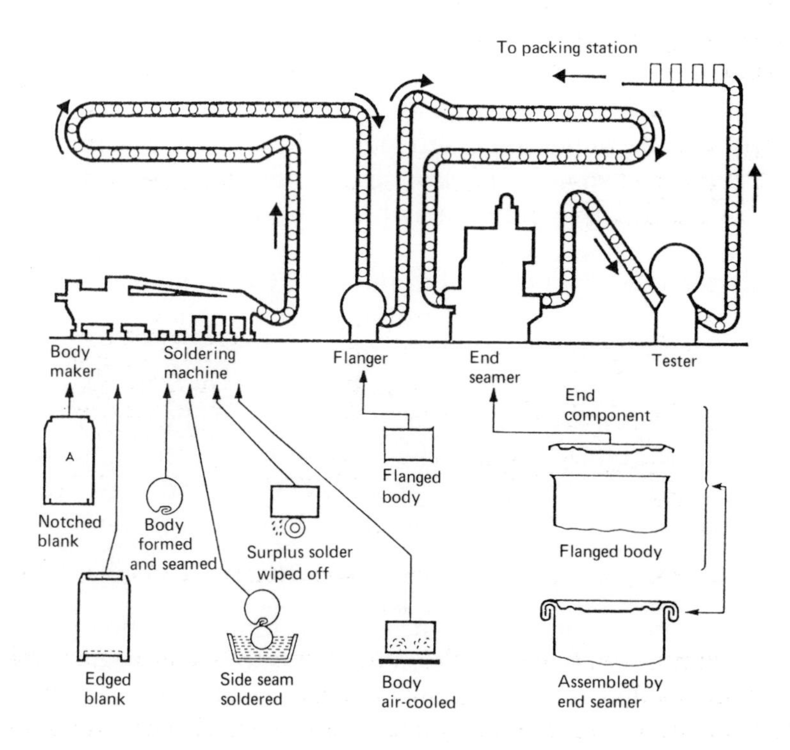

FIGURE 4.3.5 OPEN TOP CAN LINE

The stages involved in three-piece body manufacture are shown diagrammatically in Fig. 4.3.5. The first slitting operation produces strips as wide as the body circumference, including seam, and the second cuts these into blanks of the required height. Sheet edges are cut precisely square during these operations. Modern tandem slitters are fitted with carbide cutters, and are usually preset in a toolroom to achieve maximum cutting accuracy; this is essential to maintain side seam and double seam standards, together with precise lacquer stencile margins to allow efficient soldering of the side seam. The body blanks are checked for accuracy and loaded into feed hoppers of the bodymaker in which they are first passed through a rolling unit which flexes the plate to prevent flutes being formed, (related to yield point elongation, Chapter 5) and to provide an appropriate degree of bow to the blank which assists feeding through the bodymaker. Formerly softer temper grades approximating to T55 and T57 (see Chapter 3) were used. These grades called for effective breaking-down units carefully set to avoid excessive fluting occurring on the cylindrical body; although plate was temper rolled at the tinplate plant, this was sometimes inadequate to avoid fluting completely, particularly when the plate was prone to greater age-hardening. Nowadays, higher temper grades are generally specified for bodymaking, which, coupled with improved rolling units, ensure that fluting is not really a problem. The blank is then fed through and, successively:

(i) notched at the corners to reduce the ultimate side seam from four to two metal thicknesses at its extremities, to form a lap joint — this considerably helps in forming the final double seam;

(ii) the two edges parallel to the can axis are hooked to 90°;

(iii) a second hooking operation reduces their angle to about 30° to the plane of the blank;

(iv) the final operation forms the blank into a cylinder by bending it around a mandrel using a pair of wing formers; as these retract, causing the body to tend to spring back, the hooks at each edge engage to hold the cylinder; two components within this mandrel expand to hold the hooks firmly. These are then hammered flat on to a contoured spline located in the mandrel. Accurate cutting of the blank, edging it and close control at the forming station ensure a well-formed seam, essential for good soldering, a body diameter within a very narrow range and a "squarely matched" side seam for good double seaming. Immediately prior to body forming, the hooked edge is coated with a thin film of flux. If the body is to be beaded circumferentially (to increase to its resistance to abuse), incipient beads are formed across the side seam, before soldering, by a suitably designed hammer and spline.

Some types of bodymakers employ a roll former, rather than the "wing" type as is used in the welding machine described later.

The next stage, the side seamer, is in continuous motion synchronised with the bodymaker, and is fitted with a magnetic or mechanical chain conveyor to transport the body securely through the following stages:

(i) controlled preheating of the side seam to equalise expansion on soldering — a longitudinal bow interferes with satisfactory soldering (Fig. 4.3.6);

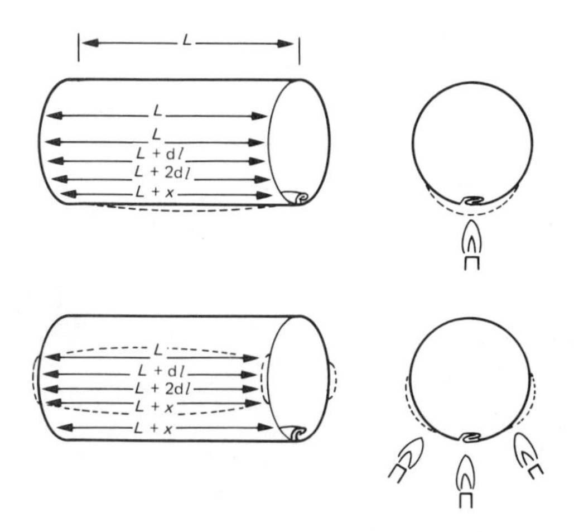

FIGURE 4.3.6 EFFECT OF PREHEAT IN PREVENTING LONGITUDINAL BOW

(ii) the side seam contacts the rotating roll carrying molten solder, causing a pool to form and by capillary action to penetrate fully into the side seam structure (Fig. 4.3.7). If the body has been prebeaded, and if it contains multiples of the final can height, finely balanced preheating and close setting of the solder roll are particularly vital.

This roll rotates in a bath of the molten solder held usually at a superheat of about 50°C, the bath being heated by gas flames. Types of solder and fluxes used are described later.

On leaving the soldering station a controlled reheating is applied to the side seam to effect good post-wiping by means of a rotating mop; this is to remove excess solder and

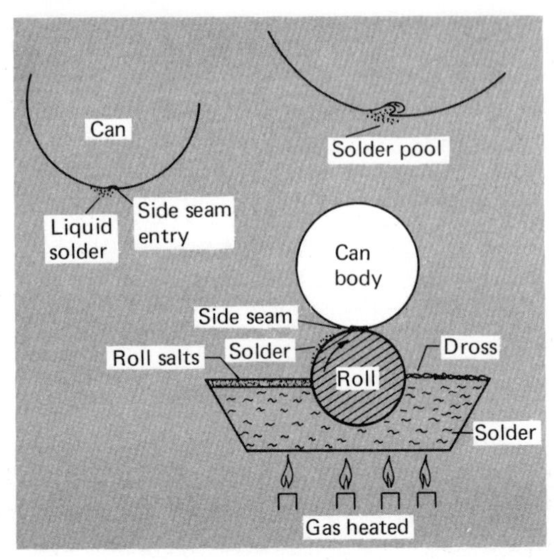

FIGURE 4.3.7 SOLDER APPLICATION (SHOWING SOLDER ROLL/CAN BODY RELATIONSHIP)

present an acceptable appearance. Both wiping conditions and temperature are critical factors in achieving a well-soldered seam with good solder fillets. Minimal metals exposure along the internal side seam region can be achieved by applying a film of suitable lacquer from a spray nozzle, which is directed on to a narrow band along the seam longitudinally. It may be applied before or after the solder operation (presolder or postsolder side striping). The interior can surface may be so treated to improve product/can compatibility; lacquer can also be sprayed externally to improve the atmospheric corrosion behaviour of the soldered seam. The lacquer is cured by the heat applied during soldering. Alternative methods have also been developed consisting of the application of resin in fine powder form, usually by electrostatic means, to the desired area, but this method is only used for welded seams at present. When a three-piece beverage can is made, the interior is sprayed with lacquer, using reciprocating nozzles and then stoved "in line".

As an alternative to roll soldering, a technique known as jet soldering is used in some areas, in which a fine stream of molten solder is applied into the side seam entrant; it requires precise control, but is said to provide a narrow attractive solder margin with minimal extraneous solder remaining over the seam area.

(iii) The side seam is then cooled rapidly by powerful air jets to avoid solder disturbance on handling at high speed.

Line speeds of up to 500 cans per minute, effectively up to 1000 cans per minute with multi-can manufacture, call for high engineering standards and close line control during initial setting and in continued running.

If multi-high, the body is then parted into its multiple lengths along the prescored lines. The two ends of each body are finally flanged outwards for subsequent double seaming on the end; the flanging operation can be by means of dies or spinning.

4.3.3 Double seaming

The principle of double seaming is shown in Fig. 4.3.8. In order to keep pace with the overall line speed, multi-headed machines must be used as the operation is complex and rather slow. It is carried out in two stages, the first operation being, after feeding the end on to the flanged body, to roll the end curl gradually inwards radially so that its flange is well tucked up underneath the body hook, to a final contour governed by the shaped roll. Seaming is completed in the second operation in which a shallower contoured roll is used to close

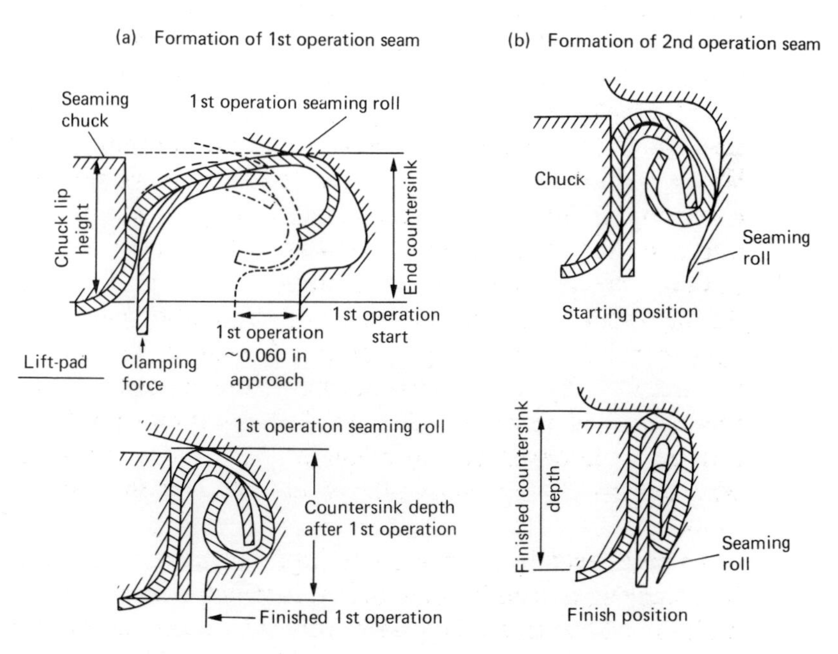

FIGURE 4.3.8 DOUBLE SEAMING

up and tighten the "first operation" partial seam to the desired tightness rating. The seam quality is defined by its length, thickness and extent of overlap of the end hook with the body hook; degree of overlap and seam tightness can be expressed in different ways, and rigid standards are laid down for the various parameters. Many canmakers have formed associations to promote the setting up of uniform component and seam dimensional standards.

Several types of multi-headed seamers have been designed; basically can and body are precisely held between a baseplate and chuck, which fits into the end; rotating or stationary chuck drives are used in which the roll-bearing units are stationary or rotating respectively. The rolls gradually move inwards under cam control, and the baseplate exerts a preset load to the can end assembly during the operation; the assembly will execute several revolutions during each of the two stages.

Modern multi-head seamers are capable of handling up to 1200 cans per minute.

FIGURE 4.3.9 A MODERN MICRO-LEAKAGE TESTER

Finally, the cans are individually tested for micro leakage under pressure in large wheel-type testers, each of which can carry up to 50 pockets per wheel. Fig. 4.3.9 illustrates a modern leakage tester.

Each machine within the three-piece canmaking line operates to a similar output level, so that they may be effectively coupled with intermediate conveyors to provide a single production entity. Each machine is normally driven independently, but with a marginal increase in speed in each successive unit; the whole production line is fully automatic so that sensors located in the runways operating as limited reservoirs will stop and start up machines as required by can flow. Similarly, if any machine fails to operate, minimum reservoirs will be maintained in front of each succeeding machine; automatic sequence restarts are provided when any failed machine restarts. Sophisticated control systems have been developed for these purposes.

4.4 SOLDERS AND FLUXES

4.4.1 Solders

Seams are soldered so as to provide:

(i) affective sealing to meet internal pressure or vacuum conditions;
(ii) a strong mechanical assembly; and
(iii) a joint having a neat appearance, free from excess solder and resistant to corrosion.

A wide range of alloys is used for soldering processes, but for soft soldering — defined as those suitable for jointing metals at temperatures generally below 400°C — the alloys of tin and lead are normally used. Occasionally, small additions of antimony, silver and cadmium are made to tin/lead mixtures. The more important solder alloys used in the United Kingdom are specified in BS 219-1977: Specification for Soft Solders.[2] This groups solders into five categories according to chemical composition, including the tin/lead series; quotes their recognised identification code; their solidus and liquidus temperature (i.e. those at which the alloy starts to melt and is just completely molten, respectively); their acceptable constituents' limits; and maxima for all likely inpurities. A further specification, BS 5625-1980: Purchasing Requirements and Methods of Test for Fluxes for Soft Soldering.[3] covers types of fluxes, fluxing efficiency, corrosion, chemical analyses, and other aspects. A related specification is BS 441-1980: Purchasing Requirements for

Flux-cored and Solid Soft Solder Wire;[4] these are only used occasionally in the canmaking industry. Similar ISO specifications are in course of preparation.

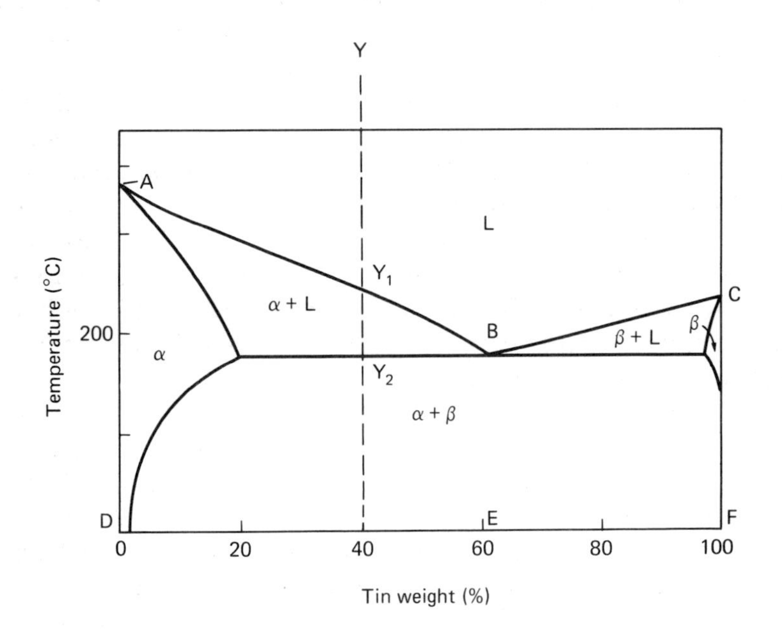

FIGURE 4.4.1 LEAD-TIN EQUILIBRIUM PHASE DIAGRAM

Binary tin/lead alloys are used for most soft soldering operations in canmaking; for some time other solders containing antimony and/or silver were favoured, as they gave stronger seams. The properties of all soft soldered joints are relatively weak in relation to the basis metal. The term soft soldering has come into general use, rather than soldering, to distinguish it from hard soldering, or silver soldering, which is now classified generally as brazing.

The development of structurally stronger seams coupled with the use of 2/98 solder (2% tin/98% lead) and later of rapid welding methods has largely replaced the need for using stronger solders; these require significantly higher bath temperatures and are more difficult to use. For many years solder compositions of 40/60 and 50/50 were the most widely used in the industry until they were replaced in many applications, by 2/98 and later with pure tin.

Fig. 4.4.1 shows the constitutional diagram of the tin/lead alloy system, from which can be gathered the solidification behaviour and phase changes that occur when a mixture of any composition is heated to well above its fully molten point and slowly cooled. In the diagram the *DEF* ordinate represents tin content from 0% to 100%

by weight. Point A is the melting point of pure lead (327°C) and C that of pure tin (232°C). If we melt together 62% of tin and 38% of lead, and follow its cooling curve, it will be found that the mixture solidifies sharply at one specific temperature, 183°C (point B); this melting point is appreciably lower than those of lead and tin. This composition is known as an eutectic ("easy melting"), and has a metallographic structure consisting of alternate lamellae of tin-rich and lead-rich phases. A mixture of 40% tin, 60% lead (Y) will be found to start solidifying at 234°C (Y_1), but will not be wholly solid until the mass has cooled to 183°C (Y_2); over this range (51°C), usually termed the "mushrange", the solder will be pasty, very liquid just below 234°C, and will gradually become thicker until it is fully solid at 183°C. For some soldering work, such as jointing lead pipes, the plumber uses this property to "wipe" the joint at an appropriate point within the "mushrange".

When solidification starts at Y_1, the solidified phase will consist of lead with a little tin in solution; with further cooling, more of the lead-rich phase will continue to separate out, and the residual liquid will gradually become richer in tin. Finally, when the metal temperature has dropped to 183°C, the final solidifying phase having become enriched to 62% tin, will consist of eutectic alloy.

The liquidus curve AB indicates the temperature at which the alloy of corresponding composition will become completely liquid when heated, and start to solidify on being cooled. Similar heating/cooling behaviour will be demonstrated as the tin content is increased beyond 62% towards 100% tin (BC).

As solders containing 38-50% tin did not possess creep strengths adequate to meet increasing demands, and the comparatively high tin contents cause a substantial increase in solder costs, development work on 2/98 solder, with a much higher liquidus point, which possessed the much needed higher creep resistance, gradually overcame the several operational problems associated with its use (negligible mush-range; much higher soldering temperatures, requirement for roller coated and side striping lacquers having much better resistance to high temperature). Later, soldering techniques were developed to make the use of pure tin — which also possesses better creep performance than solders having 40-50% tin — as a solder feasible under the demanding conditions obtaining in canmaking. Included in the development was the design of solder analysers which allowed solder composition to be checked quickly "on the line". These were based on measurement of differential cooling rates between the metals in two cells, one open to receive the bath solder, and the other, sealed, and containing pure lead.

4.4.2 Fluxes

Successful soldering of a capillary joint can be divided into four essential stages:

(i) the liquid solder must effectively wet the basis metal surface;
(ii) an adequate volume of clean fluid solder must be supplied to the seam entrant;
(iii) the solder must be drawn fully into the seam by capillary attraction;
(iv) it must alloy effectively with the basis metal surface.

Stage (i) requires the use of flux which:

(a) cleans the plate surface and prevents oxidation at soldering temperatures;
(b) does not produce residues which interfere with the formation of a continuous joint or makes unsightly deposits on its surface;
(c) has a low viscosity to allow penetration of small gaps, but is not so volatile as to disappear before the joint is made and cooled.

Thwaites[5] discusses in detail the current theories of solder wetting and emphasises that our understanding of the mechanism is still by no means complete. These aspects are also discussed by Lancaster.[6]

Complete penetration into the joint requires, in addition to good wetting, that its gap be within about 0.07-0.25 mm (0.003-0.010 in). Gap widths in the lower half of this range are preferred, but not less than the minimum, as otherwise there would be danger of flux residues or other compounds blocking solder penetration. Provided the tincoating is free from significant plating and remelting faults, is relatively free from surface contaminants, and has an adequate weight, there is usually no difficulty in machine soldering. The tincoating will quickly melt and combine with the tin in the solder. These are discussed later under test procedures.

Solder temperature must not be so high that excessive oxidation and drossing occur on the solder bath surface; an optimum temperature is usually 50-60°C above the particular solder liquidus temperature. For the solders normally used in canmaking, bath temperatures range from 300° to 370°C.

Possibly the best known fluxes are mixtures of zinc and ammonium chlorides, usually dissolved in water — sometimes other media are used — and often with a small amount of hydrochloric acid. Zinc chloride has a melting point of 285°C — rather too high on its own for soft soldering, so ammonium chloride is added to reduce the melting temperature: ammonium chloride itself is also a very effective flux, but is volatile on its own. The eutectic mixture contains 48% NH_4Cl and 52% $ZnCl_2$, but this contains far too much ammonium chloride and it will liberate large volumes of corrosive fumes during soldering: the compromise normally used in practice is 10-20% NH_4Cl. A small amount of HCl is sometimes added to prevent an undesirable reaction between $ZnCl_2$ and water. Close control of the amount of flux applied is essential, and precautions must be taken against corrosion of machine parts. As this flux is highly corrosive, it is little used nowadays in any high-speed canmaking. The principal use is for the attachment of fittings onto large containers, but even in these cases it has been superseded by spot-welding. Open top canmaking fluxes are generally based on resin — an exudate from particular types of trees; its abietic acid content, a weak acid, provides a reasonably active fluxing power; resin characteristics vary, and it is usual to specify a particular source. Resin is too viscous on its own and additions of lactic or oleic acid are made; the mixtures are dissolved in commercial isopropyl alcohol. Variants are used very widely in side seam soldering; residues are generally minimal and non-corrosive and provide a protective layer.

This type is inadequate for the more complicated designs developed to provide the much stronger side seams needed for pressurised containers (e.g. aerosols and beverage cans) and improved fluxes have been devised over recent years based on a range of chemicals; these will include Staybelite resin; amine hydrochlorides; ethanolamine compounds; sebacic acid; tin chloride; and a number of others claimed in extensive patent literature. All are effective to a degree in improving fluxing action, but are generally accompanied by a gradually increasing corrosion hazard.

Neither blackplate nor chrome/chrome oxide coated plate is solderable by any of these "non-corrosive" systems; the latter has a very resistant surface coating, similar to the oxide surface on aluminium.

Considerable effort is being directed at developing improved non-corrosive fluxes, particularly for applications in the electronic industry in which the consumption of solder is increasing. SA (synthetic activator) type fluxes under development by duPont in the United States are well-known examples; some are soluble in water, whilst others are dissolved in organic solvents. Water-based fluxes are in use in the United States, and are popular because they do not leave corrosive residues. Although a lot of this work is directed at other industries, some of it is likely to be of value to canmaking.

The British Specification BS 5625-1980[3] contains a useful systematic classification of fluxes and methods of testing desirable characteristics. An ISO version is in course of preparation.

Solder pastes have not been used to any degree in the canmaking industry, although they have been tested for attaching fittings. Flux residues can be excessive, and the need for very low oxide contents in the fine solder powder used requires stringent control.

In addition to the quoted references, the International Tin Research Institute, London, have published a number of pamphlets and papers on solders, fluxes and soldering techniques.

4.4.3 Test methods

Numerous tests for evaluating soldering systems — attempting to cover solders and fluxes as well as plate surface — have been developed over the years, with gradual improvement in their value. They can be summarised according to type as follows:

(i) *Solder spread tests:* in which a standard solder pellet is placed on a prefluxed plate held horizontally, heated until the solder melts, and after cooling, the area of spread is measured. A sophisticated form of apparatus was developed by Bailey *et al.*[7]

(ii) *Capillary rise test:* the prepared sample, consisting of clamped pairs of plate, or a single coupon bent into a triangular section, is immersed, after fluxing, into molten solder and the extent to which the solder rises within the formed sample is measured.[8],[9]

(iii) *Lap joint tests:* the specimen is prepared from two small portions of plate, fluxed and the solder applied. Assessment of joint quality is by measured lap strength in a tension testing machine.[11]

(iv) *Dynamic wetting measurements:* using purpose-built testing equipment, the solder wetting process is studied dynamically. This represents one of the most important developments in assessment of solderability. It has been developed successively by many workers; Earle[10] first published details of a viable apparatus in 1945. This was successively modified by several workers and culminated in D. MacKay's[12] design. In this, the test piece is supported on a pair of cantilever springs with the armature of a linear variable differential transformer. Output from the load cell is fed to a chart recorder. This is proportional to forces acting on the test piece, giving a graphical record of the wetting process. The MacKay design is marketed as the GEC meniscograph meter, and is now widely accepted as a valuable method of testing solderability. Photographs of the apparatus are shown in Figs. 4.4.2 (a), (b) and (c), and a line diagram of the load system in Fig. 4.4.3. A pamphlet has been issued by the manufacturer giving details of the test procedure and interpretation of the results.[14] Tests of types (i) and (iii) above are still used, but are generally useful as "weeding-out" tests only.

FIGURE 4.4.2a THE GEC MENISCOGRAPH METER

FIGURE 4.4.2b THE GEC MENISCOGRAPH METER

The meniscograph technique[14] has been used by McFarlane and Rees[11] to study interrelationships between tincoating mass, surface roughness and passivation of tinplate, and flux composition, with dilution and solder composition. An idealised test curve, showing the stages involved, is given in Fig. 4.4.4 and meniscograph traces, showing acceptable and poor wettability, in Fig. 4.4.5. These are also taken from McFarlane's paper.

SEM photographs of parted good and poor joints are shown in Fig. 4.4.6. The folds formed in the specimens are clearly seen in both photographs at ×20. At ×100, the well-soldered joint shows, after parting, typical overall "cup and cone" type fracture of the solder; in the badly soldered joint, incomplete penetration and wetting of the solder are seen. In the preliminary part of this investigation, it was demonstrated that wetting times are dependent on the type of flux used, some giving consistently low times over four different tinplate tincoating grades, and others giving a wide spread, particularly with the low tincoating grade (E2.8/E2.8). Variable results were also found between pure tin and 2/98 solders. Further work concentrated on evaluating several batches of one differential grade (D2.8/5.6). Overall, the findings have shown clearly

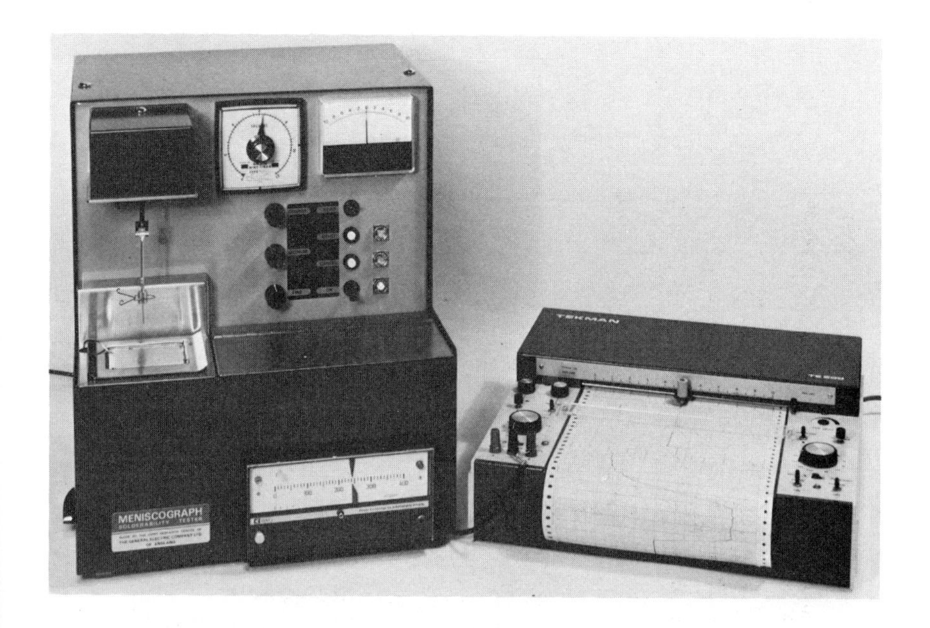

FIGURE 4.4.2c THE GEC MENISCOGRAPH METER

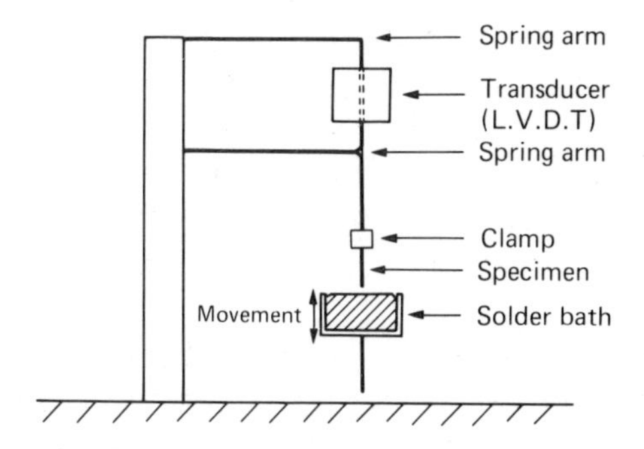

FIGURE 4.4.3 SIMPLIFIED DIAGRAM OF THE MENISCOGRAPH

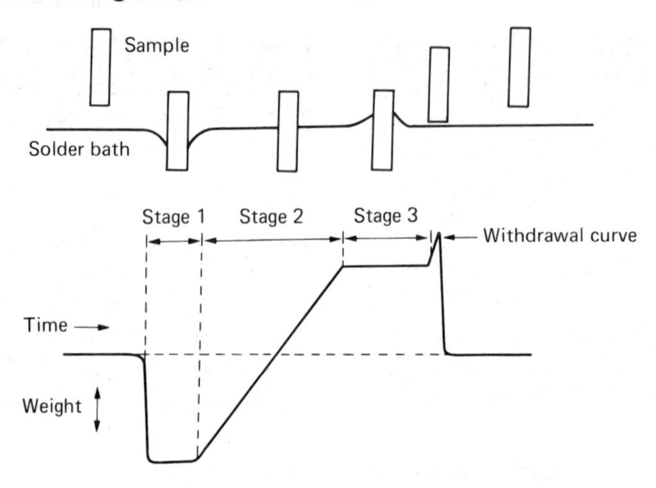

FIGURE 4.4.4 MENISCOGRAPH TEST SEQUENCE AND IDEALISED CURVE

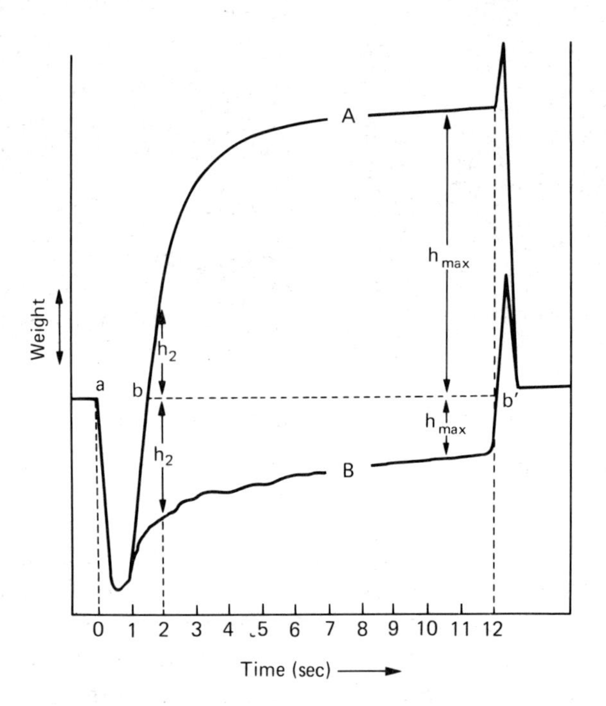

Trace A – Acceptable wettability — Wetting time (a–b) 1.5 sec
h_2 33 mm h_{max} 84 mm

Trace B – Poor wettability — Wetting time (a–b′) > 12 sec
h_2 –36 mm h_{max} 21 mm

FIGURE 4.4.5 MENISCOGRAPH TRACES

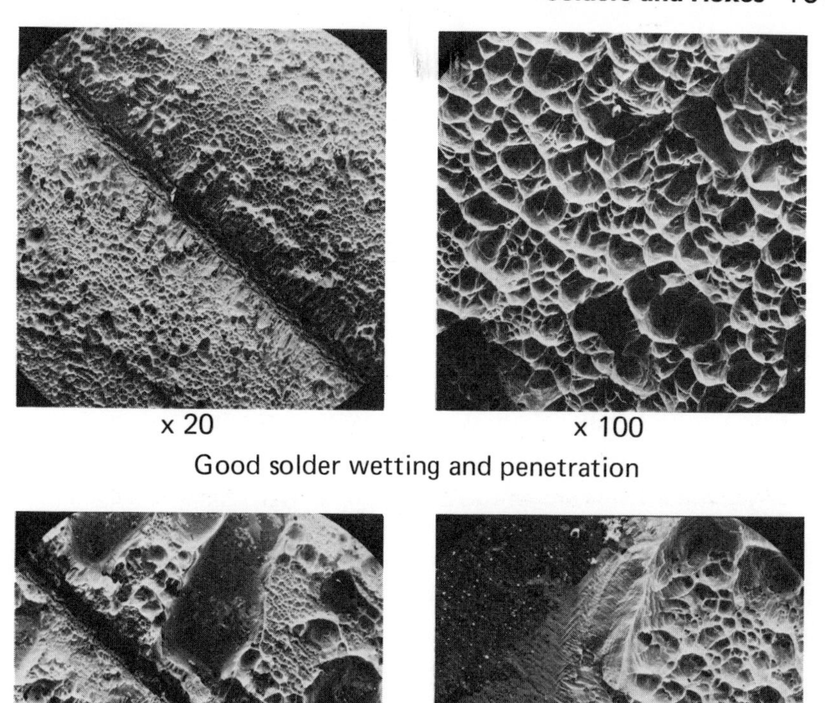

FIGURE 4.4.6 SCANNING ELECTRON MICROGRAPHS ILLUSTRATING GOOD AND POOR SOLDERABILITY

(1) The importance of flux composition and concentration.
(2) Highest joint strengths are likely to be obtained with tin-coating masses higher than E2.8/E2.8, a bright plate finish, and cathodic dichromate passivation; but results are variable.
(3) There is some evidence indicating that a "stone-finish" at the lower tincoating level is probably more difficult to solder, coupled with a flux effect.
(4) Solder composition and temperature also play a significant part in forming a satisfactory joint.

The effects of several tinplate parameters on solderability have also been studied by Coller *et al*[(13)]. Their paper reports on the relationship between tinplate solderability and surface roughness, peak count and "free" tin; solderability was measured on a GEC meniscograph, and was related to commercial canmaking solderability. Surface roughness and peak count per inch measurements were made using a Bendix QHD digital amplimeter peak counter. "Free" tin is defined as the total tin coating weight minus that in alloy ($FeSn_2$) form and as oxide. As the grade of plate most widely used for OT cans in the United States is DR 0.17 mm or 0.18 mm, enamelled over a tincoating of 2.8/2.8 g/m^2, soldered with low tin/high lead solders and a resin-oleic acid type flux, these conditions were used in laboratory testing of a large number of commercial plate batches of known solderability behaviour.

It was found that solder wettability was related to the tinplate surface characteristics according to the formula

$$\triangle W \text{ inversely as } \log \frac{RP}{T}$$

$\triangle W$ being the meniscograph wettability in mm,
R the surface roughness (μ in arithmetic average),
P peak count (number of peaks per inch), and
T the "free" tin quantity.

The term RP/T was termed the "surface factor"; this was found to correlate very closely with canmaking machine solderability. For satisfactory soldering

$$\frac{RP}{T} \leqslant 12\,000$$

the lower the surface factor, the better its solderability; thus a higher free tin value and lower surface roughness (R and/or P) are beneficial to solderability.

Fig. 4.4.7 illustrates these effects. Note that free tin is quoted in the units lb/bb (pounds per basis box of tinplate).

4.5 WELDING TECHNIQUES

Two processes were developed in about 1960 for welding side seams in place of soldering, and both have since gained wide commercial acceptance. Replacement of soldering by welding is likely to grow continuously for many years, especially where the stronger side seam will be of considerable advantage. This is particularly so where good resistance to transitory high internal pressures during post-filling operations and danger of creep failure during higher temperature storage are vital. Particular examples are

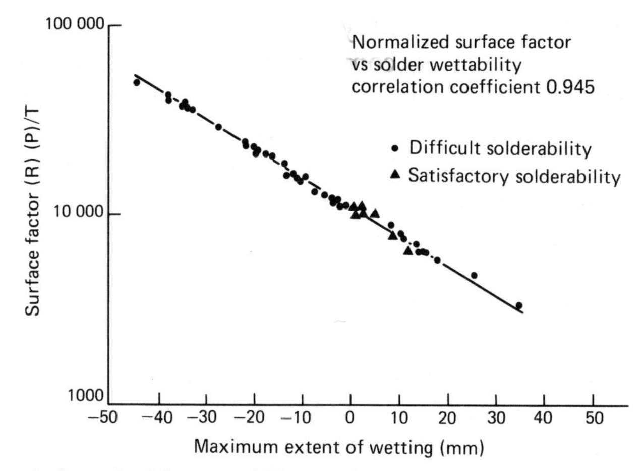

(a) correlation of solder wettability to the surface factor corrected for percent contribution. Correlation coefficient 0.945

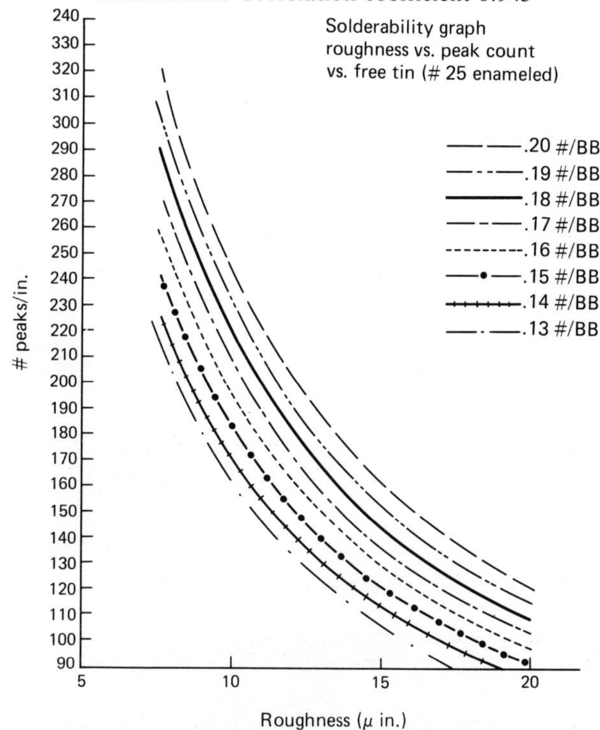

(b) solderability graph for commercial E2.8/E2.8 (#25) tinplate

FIGURE 4.4.7

aerosol containers and beverage cans, where further product development can be limited by these shortcomings. Another attractive advantage over soldering is that both welding processes offer almost "all-over" decoration: when cans are externally decorated, a prominent plain margin needs to be left at the side seam to ensure that the molten solder and its high temperature do not mar the edge of the decoration. This necessitates a comparatively wide and obvious plain margin in comparison with the very narrow (~2mm) and unobtrusive welded seam (Fig. 4.5.1.). This appearance advantage also applies to cemented side seams, but they have low strength.

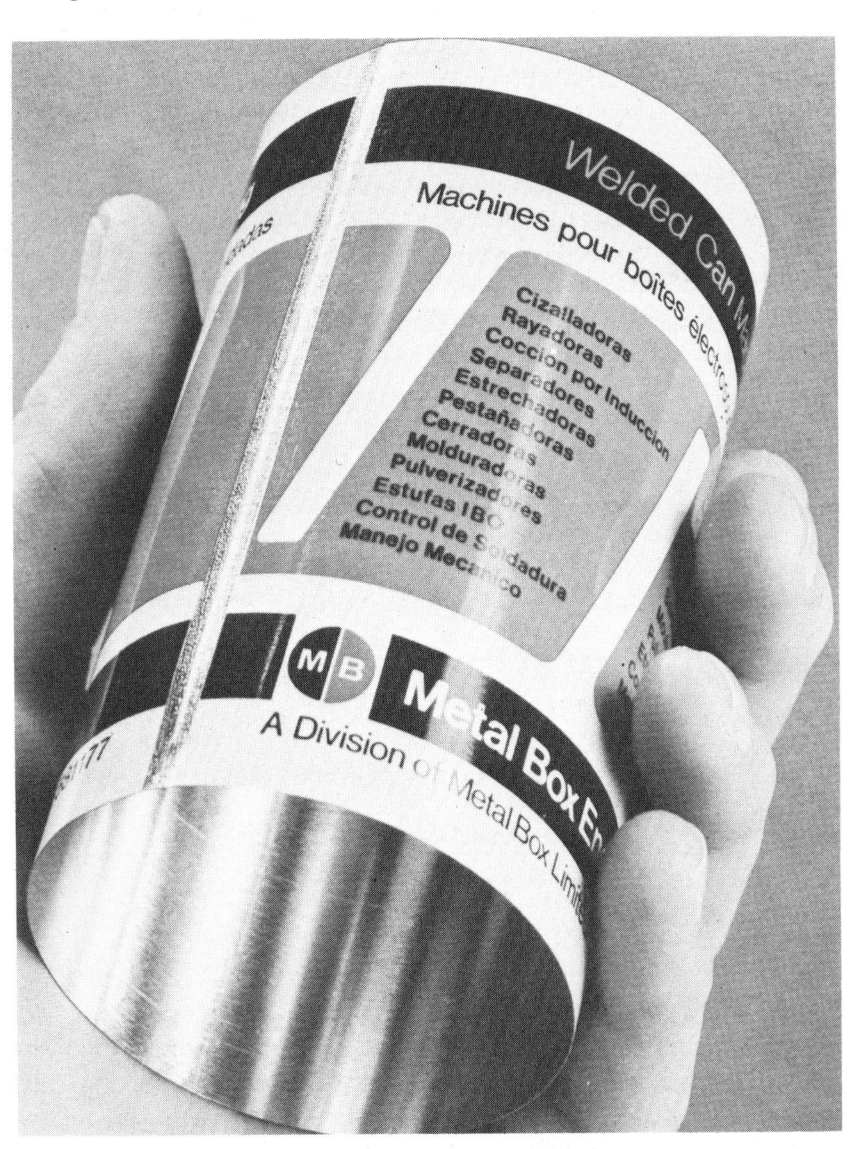

FIGURE 4.5.1 NARROW WELDED SIDE SEAM

The Conoweld technique[15] was developed by the Continental Can Corporation in the United States, initially for the production of beverage cans to be made from "TFS" plate; it has since been expanded to manufacture aerosol containers and some food cans. The process requires careful and thorough removal of the chrome/chrome oxide coating, to a width of about 2 mm along each of the blank edges which will form the lap seam, just prior to welding to ensure 100% satisfactory welds. It is vital to avoid contamination of the copper electode rolls. A specially designed square wave form of alternating current is employed to give sound, continuous, solid phase welds. High-precision body forming, with spot welding followed by seam welding, maintains an overlap of less than 1.3 mm, and the welded lap thickness is reduced by rolling to significantly less than twice the plate thickness; this is of real help towards good double seaming technology. The welding units are fitted into a modified bodymaker/side seamer assembly, and the process is operated at similar high speeds to those of modern soldering lines.

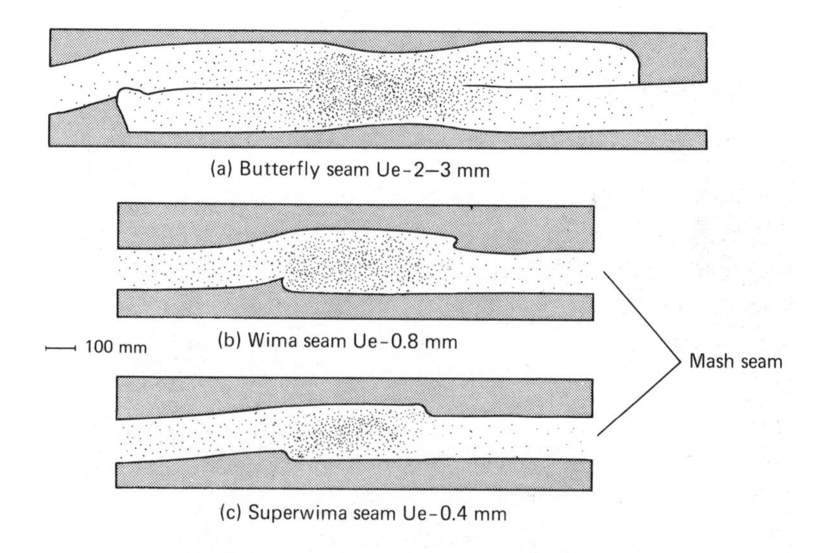

FIGURE 4.5.2 TYPES OF SOUDRONIC WELDS

Early welders developed by Soundronic AG Switzerland were for slower hand-fed production lines, making large containers. The side seams had overlaps ranging up to 4.5 mm, and the weld nuggets extended over one-quarter of the total lap width; this came to be known as the "Butterfly" weld form, and is illustrated in Fig. 4.5.2a. This was not suitable for containers to hold foods and other corrosive products, as the loose raw edges, together with some metal splashes from the welding operation, were very difficult to cover with protective lacquer. Welding speeds at that time were of the order of 15 m per minute. Since that time, considerable developments have been achieved in several respects:

(i) To improve the "Butterfly" seam structure to a solid phase weld bond having much narrower over-lap areas (reduced to a width of about 1 mm by 1975, and not more than 0.5 mm in 1978).
(ii) To increase welding speeds up to 55 m per minute in the latest machines.

These improved welds are shown in Fig. 4.5.2b and c. The improved machines, coupled with improved post-lacquering techniques, which make them suitable for producing cans for packing most beverages and foods, are described later. The combination of higher speeds and narrower overlaps require welding conditions to be controlled within narrower ranges.

The Soudronic process

To avoid the deleterious effect of electrode surface contamination on weld quality, Soudronic developed the use of copper welding rollers over which copper wire is passed (Fig. 4.5.3). The wire serves as an intermediate electrode, and is moved along with the welding of each can. As the wire is not reused, tin pick-up, interfering with subsequent welds, is avoided, thus ensuring almost constant electrical resistance. The wire is not otherwise affected,

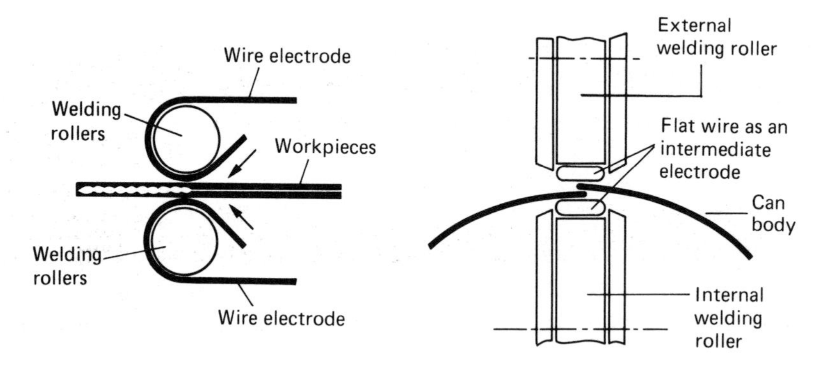

(a) Roll seam welding with wire electrodes (b) The soudronic superwima welding technique soudronic welding technique

FIGURE 4.5.3 SOUDRONIC WELDING TECHNIQUE

and is readily sold for recovery at prices around 60% of its initial cost. The Soudronic welding technique employs a sine wave alternating power supply and controls the current and the pressure applied through the electrode wheels, and the time by the machine speed. High electrical resistance in the plate overlap causes its temperature to rise rapidly, up to 1500°C, but just below that in the earlier "Butterfly" weld, the aim being to melt only the interface;

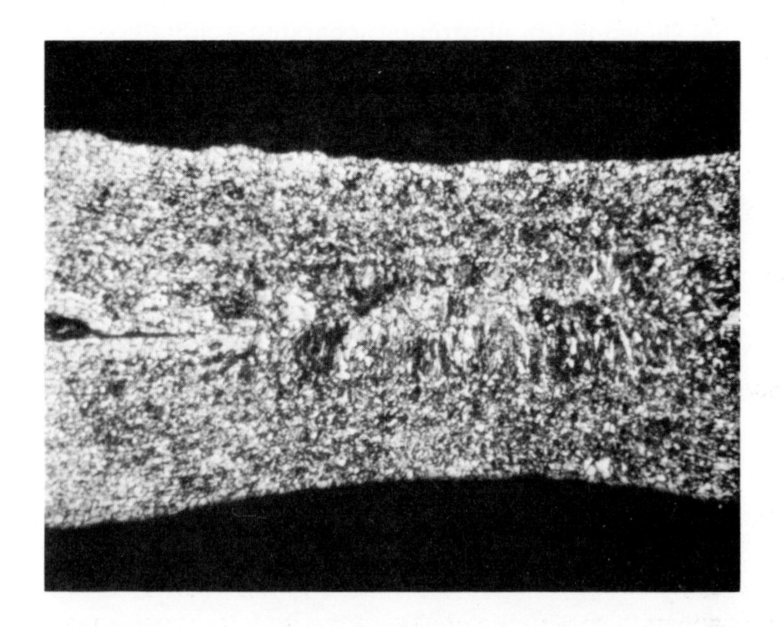

FIGURE 4.5.4a MICROSTRUCTURE OF A SATISFACTORY WELD: TRANSVERSE SECTION (x 150)

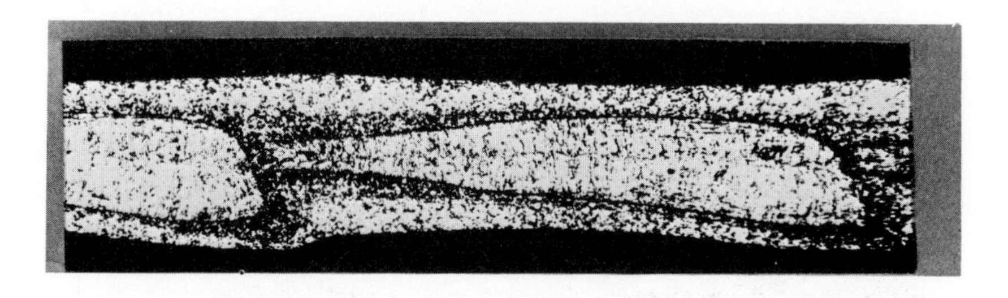

FIGURE 4.5.4b OVERLAPPING WELD NUGGETS: LONGITUDINAL SECTION (x 40)

this results in nuggets of metal with a columnar metallographic structure. Excessive heating will cause molten metal splashes to fall on to the surrounding can surface, which will be difficult to cover adequately in a post-lacquering treatment, and thus will lead to corrosion problems. Serious blow-holes can also occur within the weld nuggets, resulting in localised weak welds. Some of these conditions are shown in Fig. 4.5.4; the weld quality required will depend on the container end use, and for some non-critical applications, that shown in (c), illustrating separated weld nuggets, will be acceptable. Fig. 4.5.4e is grossly overheated, and will be accompanied by severe metal splashes and plate embrittlement.

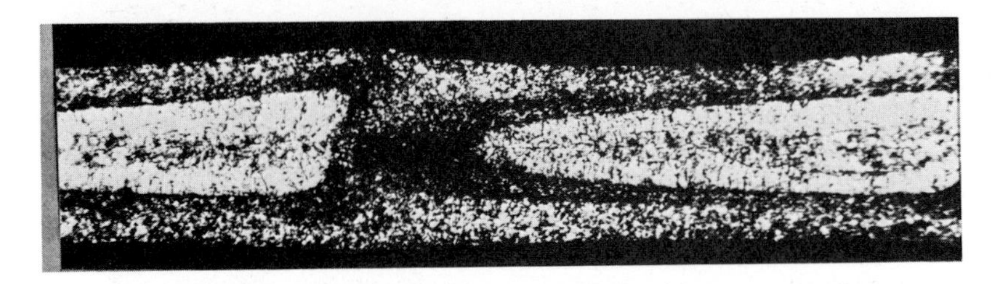

FIGURE 4.5.4c SEPARATED WELD NUGGETS (x 40)

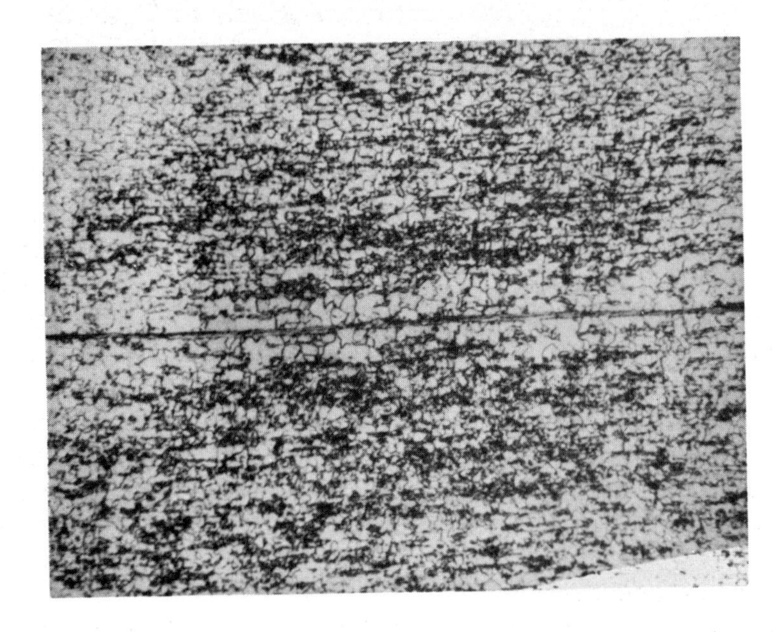

FIGURE 4.5.4d A "STUCK" WELD (x 150)

FIGURE 4.5.4e OVERHEATED WELD SHOWING BLOWHOLES (x 150)

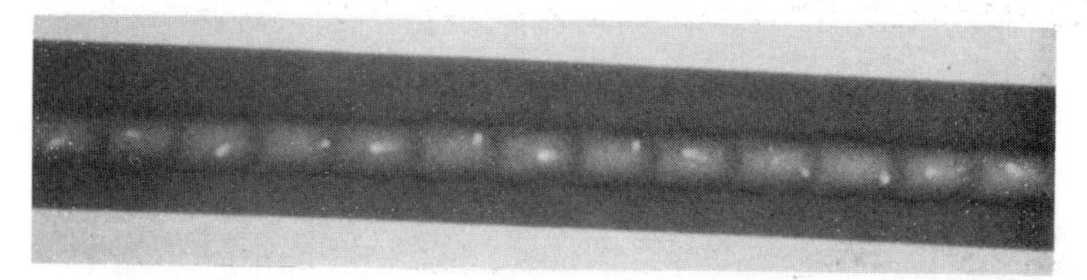

FIGURE 4.5.4f X-RAY MICROGRAPH SHOWING CAVITIES

Heating for the latest WIMA (*WI*re *MA*sh) and SuperWIMA weld forms is lower, but must always be adequate to raise the plate temperature to higher than 900°C to ensure effective solid phase bonding at all locations along the seam. The early machines used a 60 Hz power source; this was gradually raised with further development to 420 Hz in the latest SuperWIMA machines to give 850 spots per second, the distance between spots reduced to 1 mm at the maximum speed of 55 m/min. More precision was built into the Z-bar canforming station to allow accurate overlaps of not more than 1 mm and 0.5 mm respectively to be maintained; the welding pressure was also increased to assist formation of strong welds with a lower heat input. Shaerer(18) has described the technique of Soudronic welding in some detail.

Sodeik(17) has demonstrated that the total heat generated for welding arises at five different locations, as shown in Fig. 4.5.5; although the individual resistances are difficult to measure, the total resistance — the sum of all five — can be measured as a function of electrode pressure. Welding heat is generated in each of the five elements proportionately to the square of the current. Most of the total resistance was shown to originate in the steel base; variation in surface roughness, passivation or oil level has no significant effect on contact resistance nor total resistance under load.

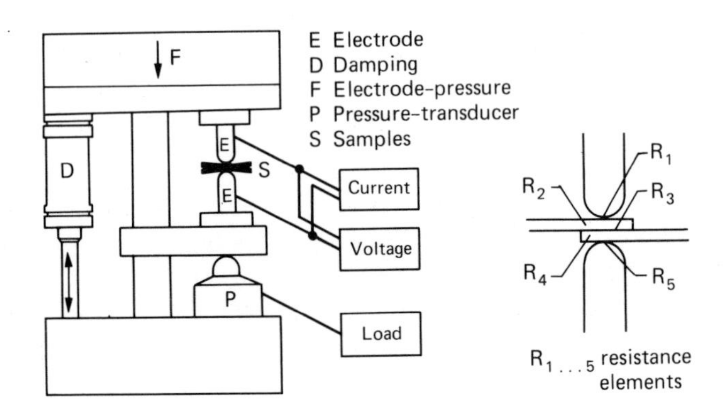

FIGURE 4.5.5 RESISTANCE MEASUREMENT

Norman[16] has described Metal Box experience in the control of welding variables with some versions of the "Butterfly" type, and concluded that the process is more sensitive to changes in welding conditions (applied current and welding pressure) that normal variation in tinplate gauge, hardness and tincoating weight. Deviations outside specified thickness or hardness tolerances, however, adversely affect weld quality, and changes in plate specification require corresponding adjustment to machine settings. For smaller can diameters (below 60 mm) temper T52 plate gives more consistent production performance. Specifying the rolling ("grain") direction to run round the can body — the so-called "C" direction — also assists flanging behaviour in the weld area. For higher temper grades, which are normally used for three-piece cans, welding conditions become more critical, and slight notching each side of the weld is sometimes adopted to reduce strain level and thereby avoid plate fracture during flanging. However, the heavily rolled double reduced plate is being adequately welded in the later WIMA machines. Williams[19] has also examined the effects of plate parameters on the weldability of Butterfly seams.

These studies have been extended by Sodeik[17] to cover the effects of various tinplate specifications on weldability in the newer machines employing higher welding current frequencies and welding speeds, and the narrower WIMA weld structures. He found that measurement of total resistance (Fig. 4.5.5) and its variations on a number of plate samples gave a useful indication of their weldability. From the results obtained with this test procedure it is concluded that conventional tinplate within the standards laid down can be welded satisfactorily in the high-speed WIMA machines, but the following special grades of plate cannot, at present, be welded:

(i) tinplate coated on one surface only;
(ii) fully alloyed, low tin coating grade (0.5/0.5 g/m^2);
(iii) blackplate, if stored for a short while (except under special conditions) but if the surface is freshly cleaned it can be successfully welded;
(iv) chrome/chrome oxide coated plate ("TFS"). (It can be welded, if freshly "edge cleaned", as in the Conoweld process.)

Improved welding techniques are continually under development. The newer and still experimental nickel coated steel (0.2 g/m^2) can be fairly readily welded, but not as effectively as conventional tinplate. The findings have been confirmed in actual welding trials carried out on the WIMA machines (ABM 250 and FBB 400). Subsequent commercial experience is in agreement with these findings.

The Soudronic welding technique has been widely adopted in most European countries for beverage and aerosol containers, and by 1980 some 5000 machines, over the twelve models available for manufacturing 45 mm diameter cans up to 200 L drums, were in use in a large number of countries throughout the world.

Welding is more difficult on the faster WIMA type machines employing narrower welds. Close control of conditions is vital, particularly electrode pressure, electrical power supply, and extent of overlapping the blank edges. These are more critical than "within specification" variations in plate characteristics; particularly, plate thickness variation within ±10% is acceptable.

FIGURE 4.5.6 RESTYLING OF WELDED CAN

The tensile strength of a good weld is equal to that of the base-plate, and, in contrast to a soldered side seam, allows severe reforming to be carried out for the process of restyling design (Fig. 4.5.6) without danger of seam failure. Frequent checks are carried out on machine settings and on weld quality; this can be done by bursting open the weld, as with a soldered side seam, by a peel test, or reversing the cylinder (Reversal test). Soudronic offer a "ball" tester which gradually deforms the weld by doming (Fig. 4.5.7a and b) rather similar to the deformation in an Erichsen or Olsen test, until the weld or plate fractures. It is hydraulically operated, and the test requires a few seconds only. A continuous monitor which checks the main weld quality parameters is available; it can function as an automatic rejector of cans not meeting set standards. Metal Box plc have announced that they also have developed a weld monitor which measures the energy employed in producing each weld nugget; it is based on a sophisticated micro-electronic system.

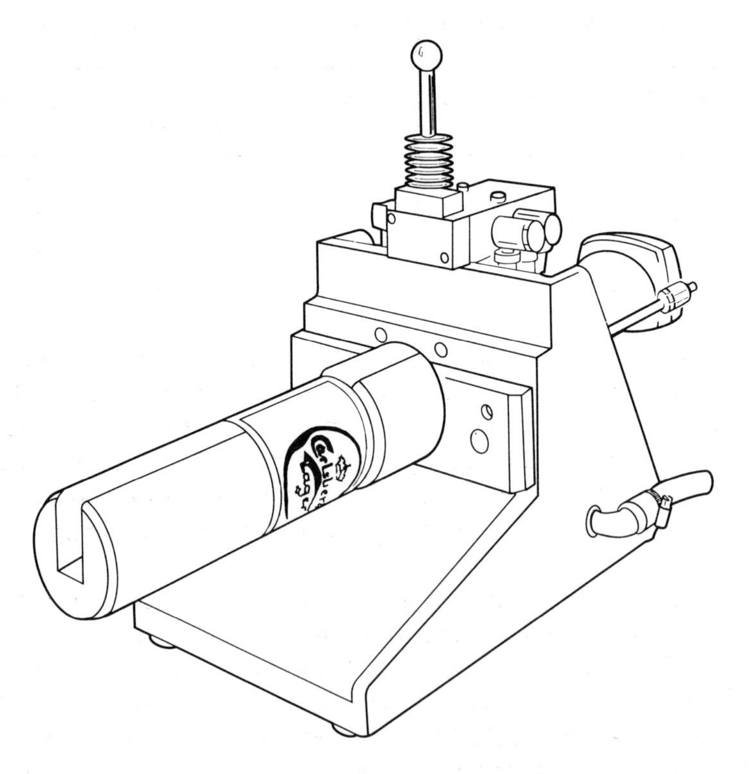

FIGURE 4.5.7a SOUDRONIC WELD TESTER

As reported earlier, chrome/chrome oxide coated steel sheet cannot be welded reliably under commercial conditions even with the edge cleaner unit, which automatically removes the whole coating, good welding on fast modern Soudronic machines is up to 20% slower than for tinplate.

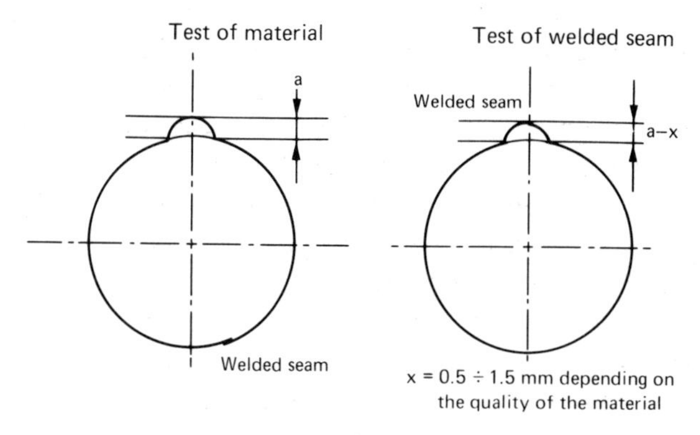

FIGURE 4.5.7b

The latest WIMA machines have the following characteristics:

Model	O/Lap mm	Frequency Hz	Welding speed m/min	Cans per min*
ABM 250	0.8	260	35	310
FBB 400	0.5	500	50	450

* of size: 73 mm ∅ × 110 mm.

The total energy demand is lower than that of side seam soldering machines; they can be switched on to operate immediately without any "warm-up" requirement; and can operate at high efficiencies, up to 97%. They can be readily integrated into existing or new can lines, and are capable of multi-high body manufacture (Fig. 4.5.8), as is also practised in can soldering processes, for the shorter height can sizes.

Changing over machines to other can sizes is claimed to be achievable within 30-60 minutes, according to diameter and height.

Although the welded seam is free from the danger of lead pick-up (which in the case of a soldered seam can only be eliminated by the use of pure tin), the weld has to be effectively coated to prevent traces of iron being picked up by some types of beverages and acidic foods. For this purpose, a driven roller coating unit (DRC) is available to be fitted beyong the welding station for side striping the seam (Fig. 4.5.9). It applies comparatively high solids lacquers to the internal weld line, and this is cured by post-solder heating.

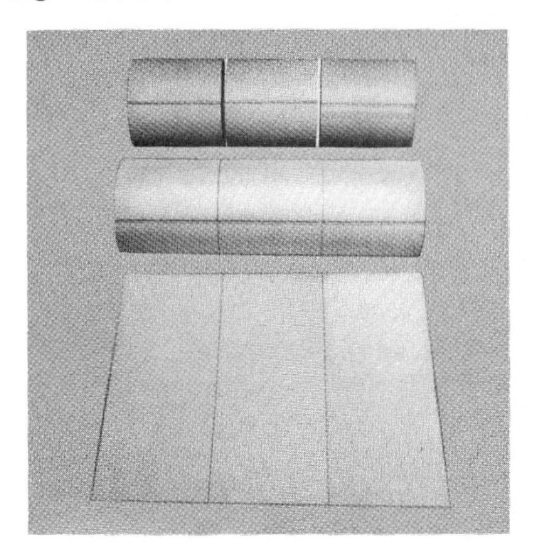

FIGURE 4.5.8 SCORING, WELDING AND SEPARATING OF MULTI-HIGH CONTAINERS

Metal exposure values, determined by the standard WACO Enamel Rater are readily obtained, within specification rates. Similarly, side stripe spraying can be carried out after welding to achieve the required protection. More recently, powder striping has been developed further to obtain cans with very high integrity.

All side stripes need to be heated for the materials to dry and cure. Both hot air and electric induction heating units are widely used for this purpose.

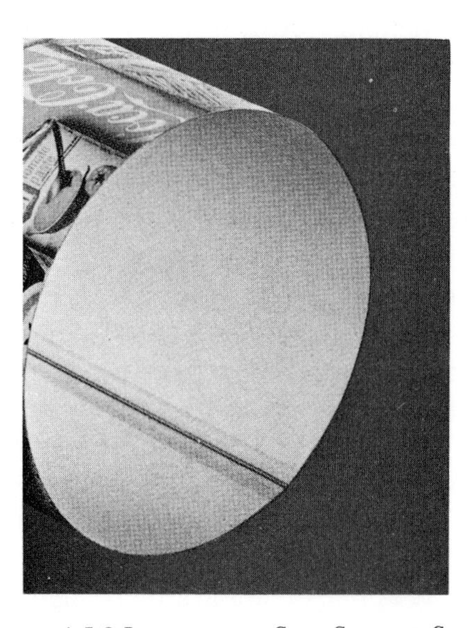

FIGURE 4.5.9 INTERNALLY SIDE STRIPED SEAM

REFERENCES

1. A.L. STUCHBERY, Engineering and Canmaking, *Proc. Inst. Mech. Engrs,*, 1965-6, 1167.
2. BS.219 – 1977: Specification for Soft Solders, British Standards Institution, London.
3. BS.5625 – 1980: Purchasing Requirements and Methods of Test for Fluxes for Soft Soldering, BSI, London.
4. BS.441 – 1980: Purchasing Requirements for Flux-cored and Solid Soft Solder Wire, BSI, London.
5. C.J. THWAITES, *Capillary Joining - Brazing and Soft Soldering Materials,* Science Research Studies Series, John Wiley, 1982.
6. J.F. LANCASTER, *The Metallurgy of Welding, Brazing and Soldering*, 2nd edn. Allen Unwin, London.
7. G.L.J. BAILEY *et al. Metal Industry,* **75,** (1949), 554.
8. M.E. FINE *et al. Trans. ASM*, **37,** (1946), 245.
9. A. LATIN, *J. Inst. Metals.* **72,** (1946), 265.
10. L.G. EARLE, *J. Inst. Metals.* **71,** (1945), 45.
11. D. McFARLANE *et al.* Second International Tinplate Conference, ITRI, 1980.
12. D. MACKAY, *Proc. Inter-Lapcan* Vol. II, p.40, Brighton 1970.
13. J.C. COLLIER *et al.* Tinplate Solderability as affected by Surface Roughness, Peak Counts and Free Tin, Second International Tinplate Conference, ITRI, 1980.
14. *The GEC Meniscograph Solderability Tester,* The General Electric Co. Ltd., Hirst Research Centre, England.
15. W.T. CHIAPPE, *Modern Packaging,* March 1970, 82.
16. G.F. NORMAN, Welding of Tinplate Containers – an Alternative to Soldering, First International Tinplate Conference, ITRI 1976.
17. M. SODEIK, Influences of Material Properties on Side Seam Welding of Cans made of Tinplate, Second International Tinplate Conference, ITRI 1980.
18. G. SCHAERER, Food and Beverage Can Manufacture by Soudronic Welding Technology, Second International Tinplate Conference, ITRI 1980.
19. N.T. WILLIAMS *et al.* High Speed Seam Welding of Tinplate Cans, *Metal Construction,* **157,** (1977), 202.

Chapter 5

Modern Canmaking Technology

5.1 INTRODUCTION – METAL DEFORMATION

Ductility — the capability to undergo extensive deformation without fracture — is probably the most important property of many metals and it is particularly valuable in modern canmaking processes. The production of sheet or strip with the requisite combination of strength and deformation characteristics involve knowledge of the response of the material to mechanical working, generally by rolling, and to heat treatment. A considerable amount of fundamental research and process development work has been carried out over the last 30 years or so. The foundations of modern knowledge date from earlier work on metal working and heat treatment, and their individual and combined effects on the mechanical properties of metals. The purpose of this section is to provide a basic understanding of the process of metal deformation and to outline how the forces required and the properties of the product can be manipulated by various methods.

Metal deformation takes place in two stages: the first stage can be described as "elastic", in that a bar or strip is stretched by an applied force, but returns to the original dimension when the force is removed. Within the limits imposed by the individual sample of metal, the extension is proportional to the force applied, i.e. the metal behaves like a spring; the general behaviour can be described more precisely by stating that the "strain", the ratio of the extension to the original length, is proportional to the "stress", the force per unit area of cross-section of the sample. When the stress applied is increased, a point will be reached when a permanent change in dimensions will persist after removal of the applied load: this is known as plastic deformation.

If we consider what is probably the simplest uniaxial stress system, that of a steel wire hanging vertically, and to the bottom end of which gradually increasing weights are applied; the additions cause the length of the wire to increase in a related manner, up to a certain limiting value. This deformation is illustrated by the portion OA of the curve shown in Fig. 5.1.1. In the region OA the relationship between stress and strain is linear, and mathematically is of the form $S = Ee$, or

$$\frac{\text{Stress } (s)}{\text{Strain } (e)} = \text{a constant } E$$

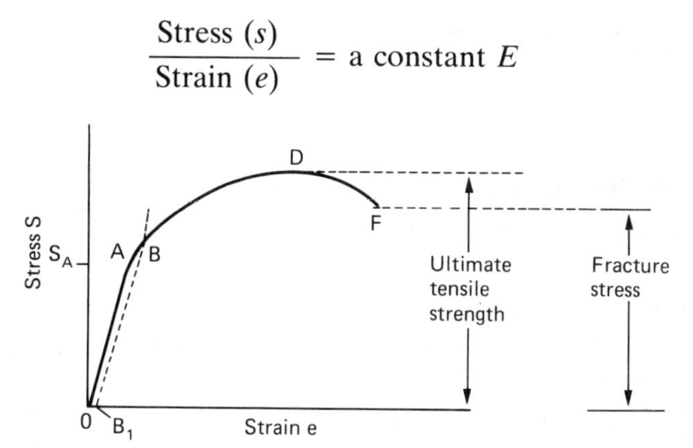

FIGURE 5.1.1 ENGINEERING STRESS-STRAIN CURVE

i.e. stress is proportional to strain; this is generally known as Hooke's law, and the constant E represents the modulus of elasticity (Young's). Point A represents the "elastic limit" or "limit of proportionality" (these are not identical, and will be defined later). Provided the applied stress does not exceed S_A, its removal will cause the strain to return along the line AO to zero, i.e. the wire will not have deformed permanently; hence the elastic deformation. It will be found, in practice that the elastic limit value is not absolute, but will vary with test conditions. As more sensitive means are used for measuring strain, the derived elastic limit will be lower. In fact it is sometimes stated that the elastic range is appreciably shorter than is often supposed; it can be taken, however, that the differences found in practice are small, and primarily of academic interest. The term "limit of proportionality" is often preferred. These terms are not always defined in precisely the same way, but the following are generally accepted:

Elastic limit — is the greatest stress the material can withstand without exhibiting any measurable permanent strain after removal of the applied load. It will be greater than the limit of proportionality, when derived using the strain measurement sensitivity usually employed in engineering studies ($10^{-4}cm^2$). Precise determination of this property requires a tedious procedure of incremental loading/unloading. Limit of proportionality is the highest stress at which stress will be proportional to strain, as observable by deviation from the straight line portion of the stress/strain curve.

If the load applied exceeds the elastic limit at A, and approaches the stress corresponding to B, it will be found on removal of the load that the relief of strain will follow the path of a line parallel to OA (BB_1 shown "broken") and the material will have suffered a permanent deformation OB_1.

5.1.1 Tensile testing

This behaviour forms the basis of a tensile (tension) test, and it is carried out in a machine capable of applying a gradually increasing load at a controlled rate — this control is vital up to the yield point stage; in most modern machines the desired controlled rate can be preset. A particular form of specimen is used which has a central parallel length and enlarged ends, the dimensions of which are controlled by Standards Institutions in the countries of use (Fig. 5.1.2).

The central portion is made parellel over a length slightly longer than the specified test (or gauge) length and the radii leading into the two wider end portions are carefully blended into smooth transition curves. The parallel portion is accurately marked with the test length specified — precise dimensions are given in all test specifications (as in Chapter 3); most call for a gauge length of 50 mm (2 in) for thin sheet metal. The increase in length occurring between the two marks, resulting from the applied load, is measured accurately by means of sensitive optical or electrical devices termed extensometers.

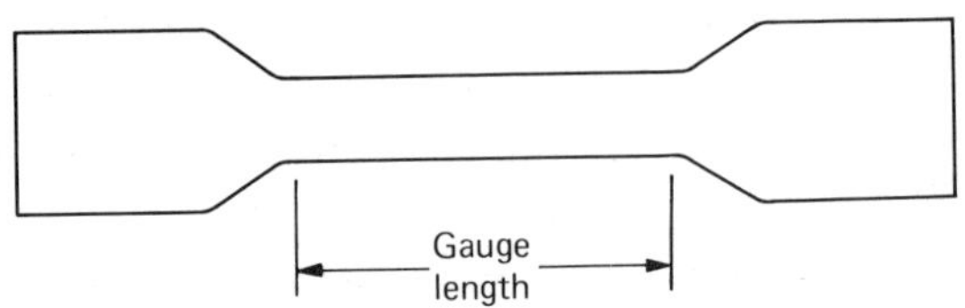

FIGURE 5.1.2 FORM OF TENSILE TEST SPECIMEN

Frequent simultaneous measurements of load and elongation enable a graph to be plotted manually. Some modern machines are fitted with precise autographic units, thereby providing a record of the stress/strain relationship automatically during the test. The test specimen shape ensures that practically all the elongation will occur within the marked length and that premature failure will not occur outside that length. The widened end portions are provided to enable them to be gripped positively in the machine clamps. These are so designed that the complete assembly and the machine must be capable of applying a truly axial load parallel to the specimen length. It is vital that the specified rate of straining is adhered to, if consistent results are to be achieved. Fig. 5.1.1 illustrates the usual type of stress/strain curve for a ductile metal and Fig. 5.1.3 that obtained on low carbon steel specimens in the fully annealed state. The latter shows that shortly beyond the elastic limit (*A*) — the onset of plastic strain — the *upper* yield point (*B*) will be revealed by a sudden drop in stress, to the *lower* yield point (*C*). The upper yield point value recorded will be dependent on the rigidity of the testing machine, and testing conditions, but it is generally between 10% and 20% higher than the lower yield point. Straining will continue irregularly at near constant load, in the form of discrete bands of deformation, generally at an angle of 45% to the direction of applied stress; these are called Luder lines (or sometimes Hartman) or stretcher strains; in "shop" parlance they are referred to as "worms". They are easily seen with the unaided eye, and as the metal thins down slightly within each band, they can also be detected by touch (Fig. 5.1.4.). The irregular localised flow will continue on further sites until the whole area is deformed to a similar extent and the bands have amalgamated into an even matt surface, at *D*. The zone represented by *CD* is termed the yield point" elongation, and in well-annealed material can approach 10% of the plastic deformation.

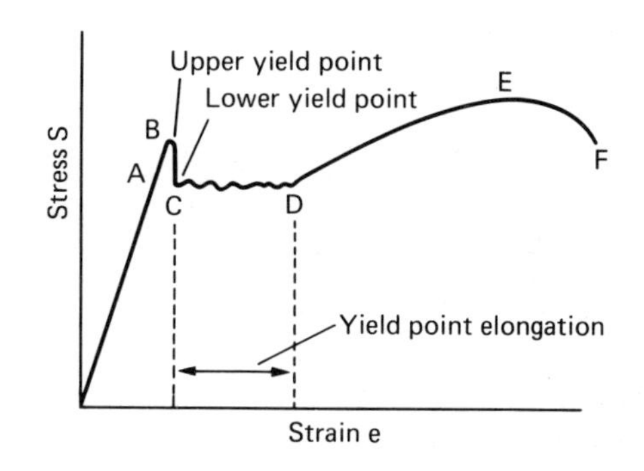

FIGURE 5.1.3 TYPICAL YIELD-POINT BEHAVIOUR OF ANNEALED LOW CARBON STEEL

As deformation continues beyong this point, the metal will elongate in a regular pattern up to *E* (the ultimate tensile strength – U.T.S.), at which significant thinning, "necking", will commence, as opposed to the thinning which is occurring continuously as straining proceeds. This will progress at an increasing rate, accompanied

FIGURE 5.1.4 STRETCHER STRAINS (LÜDER LINES) IN LOW CARBON STEEL SHEET

by a substantial decrease in recorded stress, until the specimen fractures (F). Necking will start at any point where the metal is weaker, for many reasons, than the bulk. The presence of a thin element in a pressurised container could lead to a violent rupture. The UTS value was considered in the past to be an important metal property, but it is now regarded as being of only limited value, as the point at which necking will commence.

A more realistic failure point in canmaking processes is that at which thinning is on the point of starting. The UTS is still widely used, however, as a measure of mechanical quality within a system where the steel grade is "constant" (with respect to chemical composition, rolling schedule and heat treatment etc.); any significant departure from an agreed range would be strongly indicative of non-standard material.

The drop in stress along the curve down to fracture point *F* might imply that the steel was becoming weaker as a result of this further increase in strain, but this is not so; in fact, the metal continues to strain-harden, and its strength — per unit area — continues to increase. The apparent drop in strength is purely due to the effect of the rapidly decreasing cross-sectional area in the necked region in comparison with the original.

Both stress and strain are determined on the basis of the original cross-sectional dimensions, as in the formulae:

$$\text{UTS:}\ S_u = \frac{P_{\max}}{A_o}$$

$$\text{STRAIN: } e = \frac{\Delta L}{L_o} = \frac{L - L_o}{L_o}$$

A_o is the original area, L_o the original length, L the length at fracture, P_{max} is the maximum load; e represents the average linear strain and S_u assumes that the applied stress is distributed evenly over the original area A_o but this is not generally true. This form is known as an engineering stress/strain curve. The ratio of yield stress, P_y to UTS (sometimes termed the elastic ratio) is claimed to be a useful indicator of material "stiffness". Ratios derived from published typical UTS and "yield strength" values for the usual range of tinplate temper grades show a gradual increase from about 0.60 for T50 up to nearly 0.94 for T70; double reduced grades possess ratios of up to 0.99 (neglecting the small difference due to transfer from 0.2% proof stress to yield point). However, this parameter does not appear to have been used widely up to the present. Values of the total percentage elongation of a ductile material in thin sheet form are usually more accurate than percentage reduction in area, but sometimes difficulty will be experienced in fitting the two halves of the fractured specimen back together at all accurately for measurement of total elongation. Given a comparatively short gauge length of 50 mm, that portion of the total elongation due to necking will be substantial. In an attempt to eliminate this effect, the reduction in area recorded can be converted to what is termed zero-gauge-length elongation. Accepting "constancy of volume" $AL = A_o L_o$, it can be shown that:

$$e = L\left(\frac{A_o}{A} - 1\right) = \frac{q}{1 - q}$$

where e is the elongation and q the reduction in area at failure based on a short gauge length close to the necked portion. Although this approach works fairly well with round specimens, it cannot be applied accurately to thin flat ones, due to the difficulty in measuring area.

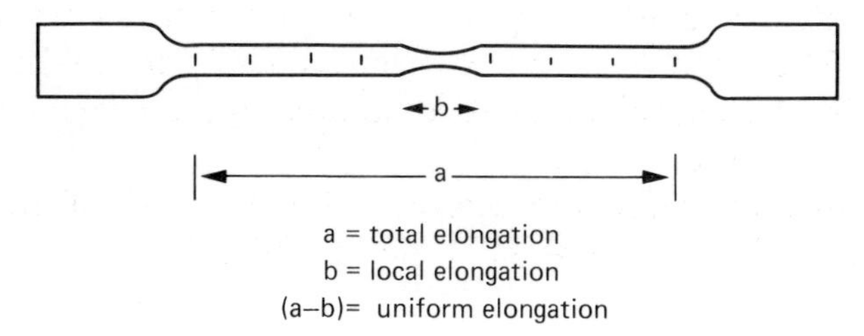

FIGURE 5.1.5 GENERAL AND LOCAL ELONGATION OF TEST PIECE

An alternative is to base percentage elongation on the uniform strain up to the vicinity of the necked part. R. Pearce[1] proposed the use of a longer test specimen (8 in parallel length) to allow distinction between uniform and local elongation.

Many attempts have been made to derive more useful data for predicting probable behaviour in "formability" of sheet metal, e.g. Warwick and Alexander;[2] none, however, have found wide acceptance, apart from the particular features of the *r* and *n* values (discussed later).

Further methods, based on more complicated procedures, have been proposed, but are not really justified in relation to canmaking activities, as tensile elongation, it is generally agreed, only correlates closely with purely stretch-forming processes.

The stress/strain curve based on original specimen dimensions suffers from several shortcomings; in particular that it does not give a true indication of metal deformation characteristics. Closer attention is currently being paid to the more realistic measures, true stress and true strain, obtained by using the instantaneous, not original, dimensions, as the latter are changing continuously throughout the test.

A further property often quoted in tensile test results is that of yield strength — or Proof Stress in the United Kingdom — which is defined as the stress required to produce a small specified amount of plastic deformation. This is usually based on an offset line drawn on the stress/strain plot parallel to the straight line portion representing elastic deformation, at the required degree of strain; amounts of strain normally specified are 0.001, or 0.002 or occasionally 0.005 (0.1%, 0.2% or 0.5%). These "off-set" yield strengths are useful for some specifications and design purposes, but they do not offer any particular merit in canmaking design.

The relationship originally used was known as the load/elongation curve, in which both load applied and resulting elongation of the test length were plotted in the measured units. As would be expected, its form was similar to that of the current engineering stress/strain curve, as each is simply applied load and resulting elongation divided by a constant (stress based on A_o and strain on original test length l_o). Thus the engineering type of relationship currently used shows the same misleading apparent drop in stress beyond the UTS zone.

All tensile tests must be carried out at a constant rate of straining, as variations can have an appreciable effect on tensile properties. Strain rate is defined as

$$\dot{e} = \frac{de}{dt}$$

for engineering, and

$$\dot{\varepsilon} = \frac{d\varepsilon}{dt}$$

for true strain.

True stress/true strain curve

As defined earlier, a stress/strain curve based on data derived from the original dimensions of the specimen must suffer from considerably limitation; it will be clear that the test piece will be elongated considerably and regularly during the test.

If the concept of "constancy of volume" is accepted — this is not fundamentally true, but is a close approximation — then the original cross-sectional area will decrease in a related manner. The false impression given by the engineering stress/strain curve, that the metal gets weaker beyond the UTS point, is simply due to the effect of the reduction in area being greater in effect than the work strengthening of the metal.

These shortcomings are avoidable if all successive incremental plots of stress and strain are based on the true instantaneous specimen length and area. A comparison between the "engineering" and true relationships is shown in Fig. 5.1.6. Differences are negligible up to yield strength, but become gradually larger with increasing strain; and the "true" curve shows the real increase in strength up to necking.

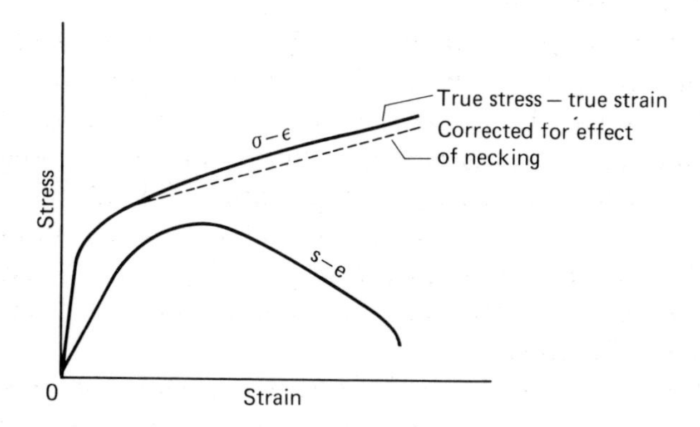

FIGURE 5.1.6 COMPARISON OF THE ENGINEERING (s-e) AND TRUE (σ-ε) STRESS-STRAIN CURVES (SCHEMATIC)

The true stress/true strain curve represents the basic plastic flow characteristics of the material and is generally known as the flow curve. It is widely used in studies of material flow.

Measurement of width and thickness during the test is difficult, but the true stress and strain values can be computed readily from the nominal/engineering values:

$$\text{Strain: } e = \frac{\Delta L}{L_o} = \frac{L - L_o}{L_o} = \frac{L}{L_o} - 1$$

$$e + 1 = \left(\frac{L}{L_o}\right)$$

$$e = \text{engineering strain;} \quad \varepsilon = \text{true strain} = \log\frac{L}{L_o} = \log(e + 1) \tag{5.1}$$

$$\text{Similarly, as true stress } \sigma = \frac{P}{A} = \frac{P}{A_o} \cdot \frac{A_o}{A}$$

(P represents the applied load)

$$\text{(Constancy of volume)} \quad \frac{A_o}{A} = \frac{L}{L_o} = e + 1$$

$$\text{Therefore } \sigma = \frac{P}{A_o}(e + 1) = s(e + 1) \tag{5.2}$$

To avoid confusion, it has become the practice to use the symbols e and s for nominal (engineering, or average) stress and strain, and σ and ε for true stress and strain respectively. In discussion of stress systems in frames and other cases, however, one still sees the former being used indiscriminately when the latter are intended.

Relationships (5.1) and (5.2) are only strictly applicable up to the UTS point; as strain becomes non-uniform due to considerable necking, direct measurement of "instantaneous" dimensions is necessary, admittedly a difficult operation especially close to the fracture point.

The flow curve can be represented by a power relationship of the form:

$$\sigma = Ke_n$$

where K and n are the strength and strain hardening coefficients.

Plotting on log/log coordinates gives a straight line relationship as illustrated in Fig. 5.1.7.

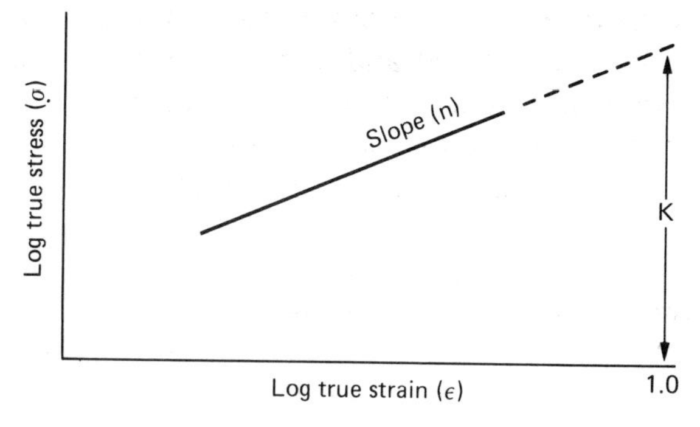

FIGURE 5.1.7 RELATIONSHIP BETWEEN TRUE STRESS (σ) AND TRUE STRAIN (ε)

5.2 DEFORMATION BEHAVIOUR OF LOW CARBON STEEL UNDER APPLIED STRESS

5.2.1 Introduction

The processes of rolling steel into sheet and strip, and of subsequent drawing into cans, will be better understood if the basic mechanisms of deformation are considered, starting with the movement of atoms within crystals or "grains", and proceeding to the change of shape of substantial polycrystalline masses.

The structure of steel which has been slowly cooled from above about 900°C consists of two phases; one is of α-iron body-centred cubic crystals with effectively no carbon in solid solution, the other is pearlite (a lamellar combination of iron carbide Fe_3C and α-iron); these phase changes are shown in the constitutional diagram, Fig. 2.5.1. If the carbon content is as low as 0.05%, the pearlite will tend to consist of globules of cementite distributed in α-iron. The α-iron phase is soft and ductile, while pearlite is strong and relatively brittle. Deformation will take place by shear in the α-iron, commencing at internal defects/dislocations, and developing along the densely populated atom ("slip") planes to grain boundaries. A single grain of α-iron will contain a very large number of unit cells having the basic structure shown in Fig. 5.2.1; this atomic structure, however, will contain numerous defects which will have significant effects on the mode of deformation. If the grain size is small, say 15 000/mm^2, it is probable that the deformation will proceed through several of them before the accumulation of structural defects prevents further progress. As a result of this, the shape can be changed, but to an extent which will be limited by the accompanying

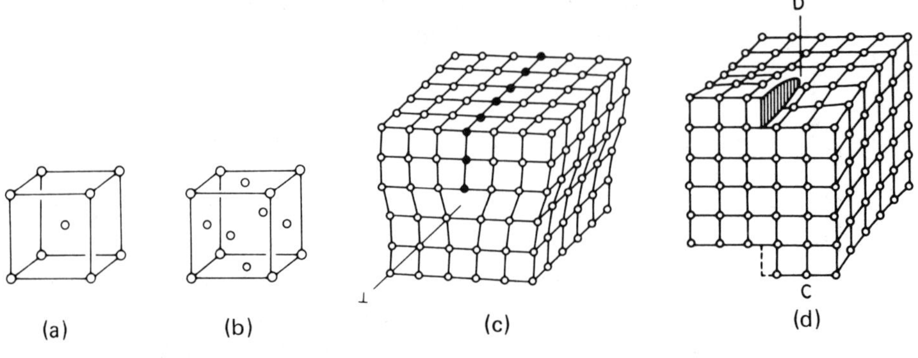

(a) Body - centred cubic structure (α - iron)
(b) Face - centred cubic structure (γ - iron)
(c) and (d) Lattice defects/dislocations

FIGURE 5.2.1 ATOMIC STRUCTURE AND DISLOCATIONS

work hardening, caused by grain distortion and increase in dislocation density. The effects of cold work on increasing the number of lattice dislocations and the beneficial effects of heat treatment are very complex and outside the scope of this book.[3,4]

If the temperature is raised to such a value that the crystals of iron are changed to the face-centred allotropic structure, above about 900°C, the ductility is increased by the increased number of slip planes, and deformation is made easier also because all of the carbon present passes into solution, thereby avoiding the occurrence of a second, harder phase. Consequently, the force required to cause slip is very much less at this elevated temperature, and any hardening due to defect generation is rapidly eliminated by atomic diffusion. Deformation can therefore be carried out to a very considerable extent without hardening and embrittlement, and at lower energy levels. The principal limiting factor in hot working is the surface damage resulting from oxidation, and it cannot be used for the production of high-quality thin sheet or strip.

The relationship between the nature and extent of deformation and the force required at any stage can be determined by carrying out a tensile test as described in the previous section, including evaluation of the r and n parameters. Low carbon steel is by no means homogeneous or isotropic, and its deformation behaviour will depend markedly on its metallographic structure, as well as on the incidence of many macro faults and inclusions, which are almost invariably present to a degree. Stress/strain relationships will be dependent in particular, on

(i) grain size; the smaller the grains, the stronger is the metal;
(ii) grain boundaries can contain second phases, subgrains and impurities to various degrees;
(iii) larger isolated second phases; e.g. cementite, NMI, etc., can be present to appreciable extents, with a resulting deleterious effect on deformation.

A non-uniform mode of deformation which is of considerable concern in drawing operations is that due to preferred orientation, or fibre texture, being somewhat similar to that of fibrous materials; this can give rise to appreciably uneven drawing, resulting in the formation of ears at the rim of the cup with associated greater thinning of the metal. This condition arises during heavy cold rolling as a result of certain crystallographic planes tending to orient themselves in a preferred manner in relation to the maximum strain direction; the non-random state can extend into the annealed material under unfavourable conditions. The tensile derived property Δr, which relates to susceptibility to earing (and described in the 5.2.2), is dependent on many processing factors; hot-mill finishing temperature, percentage cold reduction prior to annealing (generally 83-87% is preferred, but dependent on gauge), and annealing conditions. It is particularly difficult to guarantee a minimum earing quality for plate gauges less than 0.20 mm thick.

Two other non-uniform modes of deformation behaviour already referred to are the yield point elongation and strain ageing phenomena. It has been demonstrated that the heterogeneous behaviour on yielding annealed material is due to N and C in solid solution in αFe, but to avoid its occurrence both C and N contents would need to be reduced to about 0.001%, impossibly low levels in commercial practice. An alternative would be to eliminate the discontinuous yield and/or strain ageing by complete precipitation of these elements as carbides and nitrides.

As stated earlier, a few percent extension in a temper rolling operation is sufficient to suppress the effect temporarily, but it will reappear on subsequent ageing. The nitrogen content has a more important effect than carbon, probably because of its higher solubility and diffusion coefficient, and it produces a less complete precipitation during slow cooling. It has been shown that additions of carbide- and nitride-forming elements, Ti and Al, will virtually eliminate the yield point elongation characteristics even in fully annealed material.* Fig. 5.2.2 shows schematically the effects of light temper rolling and subsequent strain ageing. Thus killed steel does not show these phenomena to any significant degree, and is preferred for the more severe drawing operations.

* Titanium will combine with C and N, but aluminium will only precipitate N.

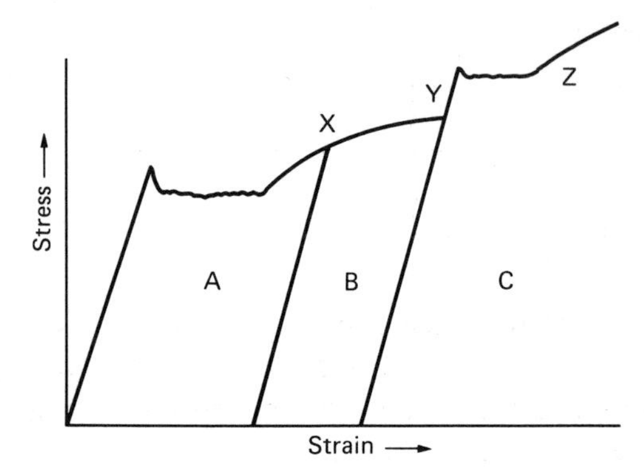

FIGURE 5.2.2 STRESS-STRAIN CURVES FOR LOW-CARBON STEEL SHOWING STRAIN AGEING. REGION A, ORIGINAL MATERIAL STRAINED THROUGH YIELD POINT. REGION B, IMMEDIATELY RETESTED AFTER REACHING POINT X. REGION C, REAPPEARANCE AND INCREASE IN YIELD POINT AFTER AGEING AT 300°F.

Very close control of all processing conditions, from steelmaking through to the final stage, is essential to avoid the incidence of these faults to an excessive level.

5.2.2 Determination of the plastic strain ratio *r*

This plastic strain parameter indicates the ability of a sheet metal to resist thinning or thickening when subjected to either tensile or compressive forces in the plane of the sheet. It gives a measure of plastic anisotropy and is related to the preferred crystallographic orientations within a polycrystalline metal. This ratio is therefore a significant measure of drawability in the deep drawing of cylindrical flat bottom cups, where a substantial proportion of the blank is drawn from beneath the blank holder into the die opening. As with most other metals, low carbon steel sheet will have different *r* values in different orientations relative to the rolling direction; it is a general practice, therefore, to test specimens taken at 0°, 45° and 90° to the rolling direction.

The parameter is defined as $r = \dfrac{\varepsilon_w}{\varepsilon_t}$

(where ε_w = width strain, and ε_t = thickness strain)

As measuring changes in thickness with sufficient accuracy can be difficult, the change in length may be measured instead.

Through constancy of volume: $w_o t_o l_o = w_f l_f t_f$

thus $\dfrac{t_o}{t_f} = \dfrac{w_f l_f}{w_o l_o}$

Therefore $r = \dfrac{\ln (w_o/w_f)}{\ln(\dfrac{w_f l_f}{w_o l_o})}$

The test is carried out in a tensile testing machine using the standard specimen described earlier; as a significant yield point elongation would result in inaccurate values, the specimen will be strained to beyond that zone but to less than the UTS.

As the level of strain can have a significant effect on the r value in plate having inherently high values (especially in the AKBA grade), it is the practice for these applications to base measurement on a tensile strain in the range 10-12%. The decrease in r value linearly with increasing strain is discussed by Jenkins(5) and by Hu.(6)

Several determinations are normally made for each plate sample, and the r value derived is significant to 0.01.

When testing at three angles to the rolling direction, the weighted average of the ratio is given by:

$$r_m = (r_o + 2r_{45} + r_{90})/4$$

The earing tendency of the material is related to its planar anisotropy; it is calculated from the coefficient

$$\Delta r = (r_o + r_{90} - 2r_{45})/2$$

As in the case of a standard tensile test, accurate specimen preparation is vital, and all test conditions laid down must be adhered to, especially the straining rate.

Amongst others, a standard test method for determining r is given in the ASTM Standard E577 – 74 included in the 1981 *Annual Book of ASTM Standards*.

5.2.3 Determination of tensile strain-hardening exponents (*n*-values)

This exponent is also derived using stress/strain data obtained in a standard tensile test. These data can be utilised because stress and strain follow the power curve relationship

$$\sigma = K\varepsilon^n$$

where σ = true stress $\quad$ K = strength coefficient
ε = true plastic strain $\quad$ n = strain hardening exponent

over the range of interest of the true-stress versus true-strain curve. It estimates the strain at the onset of necking in a tensile test, and thereby provides an empirical parameter for judging relative stretch formability between samples. It also indicates the increase in strength of the material due to plastic deformation. Although the exponent is useful for forming operations involving an appreciable amount of metal stretching, it is not regarded in general as being closely related to drawing of deep cylindrical cups.

The tensile test carried out to provide the required data must conform to the test conditions laid down, especially in respect of strain rate. Observations of load and corresponding strain are made at several points during straining, the first immediately after the yield point elongation, and the last just prior to the point at which the UTS occurs.

The strain-hardening component is usually determined from the logarithmic form of the representative power curve within the plastic range as follows:

$$\log \sigma = \log K + n \log \varepsilon$$

True stress and true strain are given by:

True stress $\sigma = s\ (1+e)$
True strain $\varepsilon = \ln\ (1+e)$

where (σ,ε) are the true stress versus true strain pair at the selected points;

s = engineering stress, $= \dfrac{L_I}{A_0}$

e = engineering strain, $= \dfrac{L_I - L_o}{L_o}$

The standard* includes methods of computing strength coefficient, K: and other factors.

Sophisticated instruments are now commercially available for the rapid determination of both r and n values; these are based on tensile testing machines and incorporate minicomputers. Precise specimen measuring gauges and graphical recorders can also be fitted. They can provide data for dimensional changes after set strains, e.g. 10% and 18% elongation, and these together with the normal stress/strain values, provide rapid and accurate means of comprehensive sheet metal testing.

5.3 DEEP DRAWING

The process of deep drawing has been widely used for a considerable time in the metal-forming industry; and the canmaking industry has produced comparatively shallow cylindrical cans for fish and meat, and non-food products for a number of years. However, as an engineering process, it has been far in advance of theoretical understanding of the principles involved; over the last 30 years or so, however, considerable in-depth study has been made of the process in many countries, and the industrial art has been changed to a well-established applied science. These efforts have resulted in substantial improvements to the quality of sheet metal, tool design and quality, lubrication techniques, etc., and to sophisticated high-speed presses. Deep drawing is used for producing a wide range and size of components. This introduction to the process will be related to the manufacture of cylindrical cups with a flat base, by means of a cylindrical punch and die set assembled in a power press.

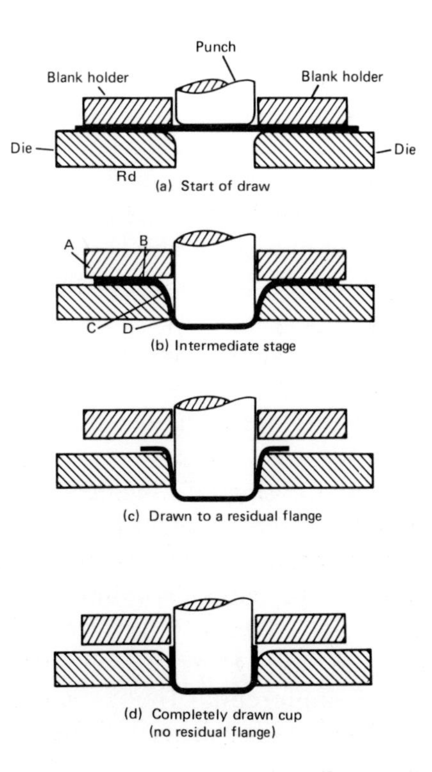

FIGURE 5.3.1 DRAWING OF A CYLINDRICAL CUP

Figure 5.3.1 illustrates schematically a section of typical tooling, and Figure 5.3.2 the metal flow taking place in the drawing operation. As the tool assembly descends, a circular cutter forms a blank which is then held down against the die by a cylindrical blankholder. Further downward travel causes the punch to effect the following processes:

(i) The peripheral flange *AB* is drawn inwards radially over the face of the die, wrinkling due to the gradually decreasing circumference being prevented by the pressure exerted by the blankholder. At the same time the central portion of the blank is stretched around the end of the punch, to form the base of the can (given as stage (iv) below).
(ii) The flange material at *BC* is first bent around the die radius R_d, and is then unbent to form the vertical cup wall *CD*.
(iii) The cup wall is stretched and thinned under tension.
(iv) The cup base *EF* is stretched over the face and around the radius of the punch.

Although the principal forces applied to the blank are tensile, considerable indirect compressive forces result from the reaction of the workpiece with the die and there are substantial frictional forces due to the blankholder pressure, and to the movement of metal between the punch and die. This complex system of forces has been the subject of theoretical and practical investigation dating from the classic work of Swift in the 1950s. More recent mathematical analyses are based on improved methods of measurement of the forces involved and on better knowledge of the response of metals to applied forces. A summary of Swift's work has been prepared by Willis,[(7)] and extensively reviewed by Alexander.[(8)] Duncan and Johnson[(9)] have provided a simplified general account of deep drawing, and valuable test are available by Dieter[(4)] and Blazynski.[(10)]

Figure 5.3.2 shows a "pie-shaped" segment of a cup during the drawing; it illustrates the three different types of deformation that take place. As a result of the metal in the flange *AB* being drawn inwards (stage (i) above), the outer circumference decreases continuously from its original πD_o, to that of the formed cup πD_f, or if fully drawn without a residual flange to πD_p. That area of metal is therefore subjected to a circumferential (hoop) compressive stress, in addition to the radial tensile stress. Wrinkling of the flange material, which would normally occur as it is drawn inwards, must be prevented, and this is achieved by applying just adequate pressure on to the draw ring (the blankholder). Application of

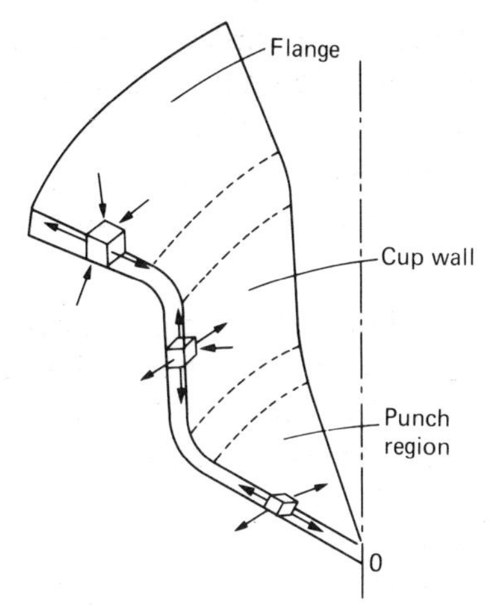

FIGURE 5.3.2 STRESSES AND DEFORMATION IN A SECTION FROM A DRAWN CUP

excessive pressure will increase the drawing load and therefore the chance of metal failure. Two methods of applying pressure are normally used (i) a constant clearance (plane strain), and (ii) constant pressure (plane stress) system; these are discussed by Alexander,[8] Blazysnki[10] and Johnson and Mellor.[11] As the metal volume must remain sensibly constant the flange thickness increases as it is drawn inwards; the extent is dependent on the material properties, blank diameter and punch profile radius, together with a circumferential variation in thickness increase according to the Δr value of the material. The thickness will be modified by thinning at later stages in the draw, and of course, will also depend on tool accuracy and the setting of the press.

Bending of the flange material over the die radius R_d under tension will cause thinning — its extent being dependent on the size of the radius, the radial tension and especially on the ratio R_d/t (t = plate thickness); and also on initial material properties and the modifying effects of cold work resulting from drawing it inwards radially over AB; in a similar manner, further thinning occurs when unbending takes place to form the cup wall. The principal stress in the cup wall is tensile in the direction of the punch load, but with some hoop tension. Any further stretching and thinning in this zone will be dependent on punch load and extent of work-hardening of the material. The combined thinning will modify the extent of thickening which had taken place on drawing in the flange. There is a narrow zone between the punch and die radii which is not bent at any stage.

It is thus possible to deduce that thickening will occur at the zones where radial tension is less than hoop compression; the converse also follows, that a higher radial tension will probably cause thinning to take place. Experience confirms that thickening is usual near the rim, and thinning nearer the die throat; this difference in stress will determine the plastic deformation, and the rupture condition will relate to the tensile stress value. The weakest point will occur at the inner radius shortly after the start of drawing, when the radial stress normally is greatest.

In conventional deep drawing it is essential to ensure that clearance between punch and die allow for the thickening which may occur in drawing, if ironing of the material — squeezing between punch and die — is to be avoided. Generally, clearance should be not less that 10% of the material thickness, and a value up to 40% may be used in practice.

The punch force required to form a cup is the sum of several factors:

(i) the ideal force of deformation, i.e. that required purely to cause the theoretical deformation,

(ii) all frictional forces, plus any needed for non-optimum deformation; and, if operating,

(iii) the force needed for any ironing of the metal.

These are shown diagrammatically in Fig. 5.3.3. Ideal deformation force increases continuously throughout drawing because the strain increases; and the flow stress also, as a result of strain hardening. The reduction in frictional forces is a result mainly of the reducing area of the flange under the draw ring; resistance to drawing accross the draw ring is a major cause of overall friction. Any ironing force required will be highest as a result of the maximum thickening of the flange towards the end of drawing. The difference between total drawing force and that required for ideal deformation, bending followed by unbending, non-optimum tool design, frictional stresses, etc., is strictly not needed and is referred to as redundant or inhomogeneous straining.

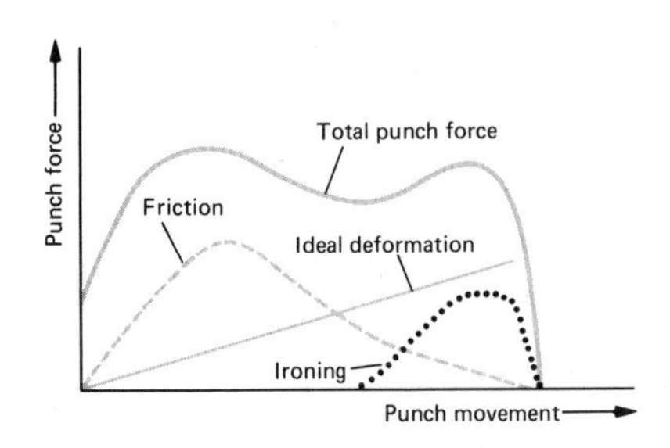

FIGURE 5.3.3 PUNCH FORCE VS. PUNCH MOVEMENT (SCHEMATIC)

Swift[(12)] calculated for a particular case that 70% of the total work was required for the raidal drawing of the metal, 13% was required to overcome friction, and 17% for the bending and unbending around the die radius.

Sachs[(13)] carried out an analysis of the forces in equilibrium during the formation of a deep-drawn cup, and concluded that the total punch load was given approximately by the function

$$P = [\pi D_p h(1.1\sigma_o) \ln \frac{D_o}{D_p} + \mu\,(2H\frac{D_p}{D_o})]\, e^{(H\pi/2)} + B$$

in which σ_o is the average flow stress,
D_p the punch diameter,
D_o the blank diameter,
H the hold-down force,
B the force required to bend and restraighten,
h the average wall thickness (not identical to t, initial plate thickness), and
μ the coefficient of friction.

The first term of this equation expresses the ideal force needed to form the cup; the second gives the friction force due to the draw ring (blankholder). More detailed analyses have been presented later by many workers (Further Reading).

In the deep drawing operation, the load is applied initially by the punch to the base of the cup, and then transmitted to the cup wall. As the narrow band of metal in the cup wall just above the punch radius has undergone only minimal tensile strain, and has therefore not been significantly strengthened, metal failure is most likely to occur within this zone by necking and tearing. The stress required to cause failure will be roughly equal to the initial tensile strength, but

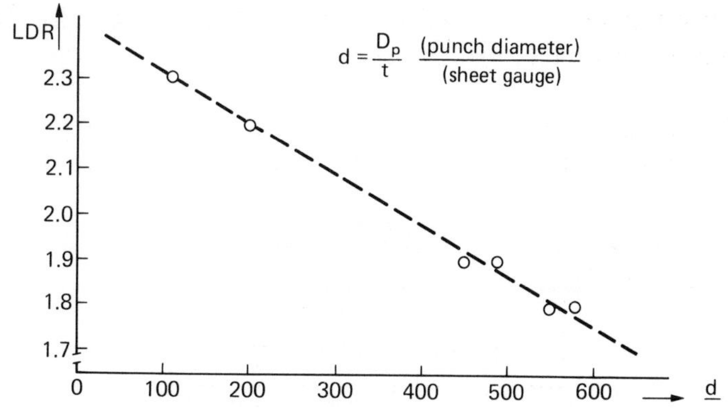

FIGURE 5.3.4 RELATIONSHIP OF LIMITING DRAW RATIO WITH RATIO OF BLANK DIAMETER TO MATERIAL THICKNESS $\frac{D_p}{t}$

increased by that due to plane strain. The plate's ability to strain-harden, its n value, does not therefore play a major part in this process.

Drawability is expressed, for this process, as the limiting drawing ratio (LDR or β_{max}) given by the ratio of the blank diameter to that of the drawn cup; it will vary according to the punch diameter and plate thickness expressed by the ratio D_p/t. Figure 5.3.4, due to Siewert and Sodiek[14] shows this relationship with tinplate for D_p/t ratios up to 600. LDR values of up to 2.3 are possible under optimum conditions, including good deep drawing quality plate and lubrication conditions; this will equate, for a popular DWI first operation cupping having a cup diameter of 90 mm and plate thickness 0.30 mm, to an LDR of 2.05. As the cup diameter is increased, and/or plate thickness is decreased, the LDR will be reduced. For some cases of DRD practice, where an initial cup diameter of 100 mm and plate thickness of 0.18 mm may be used, the LDR will drop to about 1.8 (D_p/t being ~ 550), dependent of plate grade and lubrication conditions. LDR values used in practice will be lower than the optimum to allow for variability, including that due to tool wear, etc.

Several of the tool parameters are particularly important in regard to the drawing behaviour. To obtain good control of metal flow between punch and die, the die radius (Dr) is usually specified to lie between $5t$ and $10t$*; for a flat faced punch, its radius is usually $8t$, but recent work appears to recommend that up to $15t$ is preferable.

As mentioned earlier, the blank holding load should be closely controlled to that required, merely to prevent the formation of wrinkles in the flange area. It is often expressed as a percentage of the basic drawing load (not the total press load); on this basis the percent value will increase as the blank thickness is decreased, and of course the absolute drawing load will decrease. For a low-carbon steel blank of thickness ~ 0.5 mm, the BH load will be of the order of 45% of the drawing load. These and other data are given by Gary.[15] Some general comments are given later on tool design and precision manufacture, coupled with types of presses used.

5.4 DRAWING AND WALL-IRONING (DWI) PROCESS

This was the first of the two modern canmaking processes to be developed, based on presswork only, and was aimed at the production of cans having a large height:diameter radio for pressurised beverages, The technique has been continually improved over the

* t = plate thickness

last 15 years, and has now become a widely established production method. It gives several advantages over the soldered three-piece can:

(i) more effective double seaming of the top end on to the body, as it does not contain the sensitive side seam junction, particularly valuable for processed foods and carbonated beverages packed to a high internal pressure;
(ii) significantly lower metal usage and cost;
(iii) more attractive appearance due to the absence of a side seam;
(iv) some ecological advantages.

These developments incidentally allowed large-scale use of low aluminium alloy plate as a serious competitor to tinplate, even though it cannot be soldered or welded at high speeds. The alloy used — 3004 H19 — has been found to be particularly useful, and in fact the earlier developments of the process were carried out using this material, rather than low carbon steel.

The main applications of aluminium to date have been to the beer and soft drinks products, but some cans for processed foods are also produced.

The technology requires a circular blank for forming the cylindrical body, and internal post-lacquering, with usually external decoration for beverage cans. Use of metal in coil form is almost universal (in a few cases scrolled sheet is being employed). The highly automated process produces up to 1000 cans per minute, with closely controlled linked systems; modern installations are planned on multi-lines capable of producing 1 billion cans per year, equivalent to 2 million per day. The system consists of the following basic stages, as illustrated in Fig. 5.4.1.

1. The strip is fed from an uncoiler onto an inspection/lubricating unit, which checks the material for pinholes, off-gauge portions, etc. to ensure minimum down-time and no danger of damage to expensive tooling, and then applies a controlled amount of the drawing lubricant, usually by a roll applicator.
2. A first-stage cup is drawn in a heavy duty high-precision double-action press fitted with combination tool (i.e. for blanking and drawing), multiple die — 6, 7 or even 8 cups may be produced in one press stroke staggered across the strip. The cups are fully drawn through the die, i.e. without a residual flange. By appropriate design and precision toolroom practices, tool interchangeability is normally possible. Pillars are incorporated into the press design to ensure precise tool alignment.

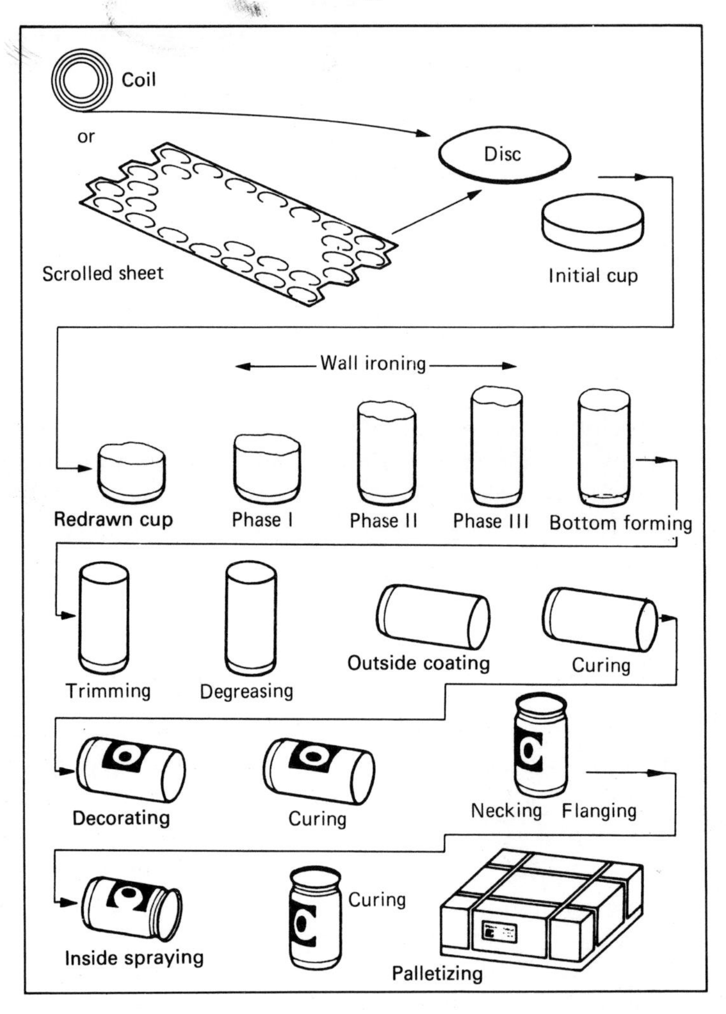

FIGURE 5.4.1 DRAWING AND WALL-IRONING OPERATION SEQUENCE

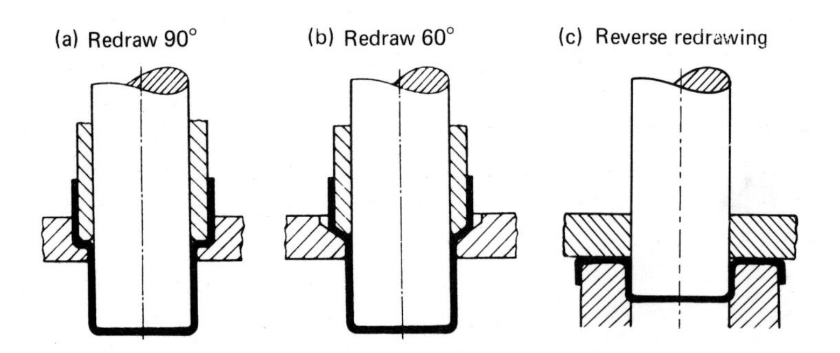

FIGURE 5.4.2 REDRAWING TOOL DESIGNS

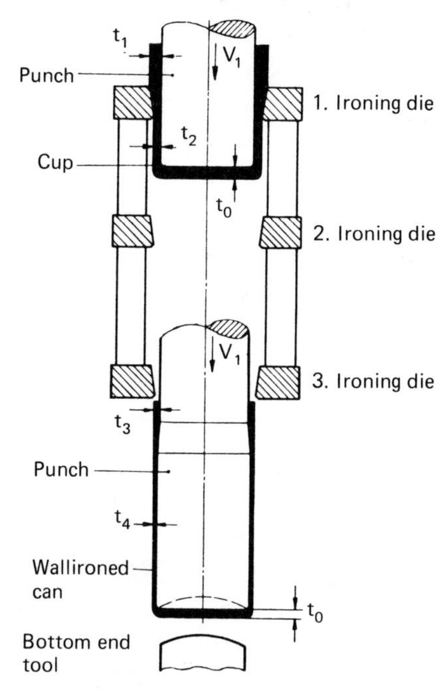

FIGURE 5.4.3 WALL-IRONING WITH 3 IRONING DIES

3. The cup is then redrawn to the final can diameter either in a separate double- or single-action press, or integrally at the beginning of the next stage, using the type of tooling shown in Fig. 5.4.2. The first stage cup must be located precisely in the redraw tool to ensure even drawing; this is achieved by means of a cup locator. Cups are removed from the punch return stroke by means of a "sprung fingers" stripper.
4. Wall ironing, in which the redrawn cup wall thickness is reduced from its original gauge of 0.30 mm to 0.10 – 0.14 mm, normally by means of three ironing rings, of the form shown in Fig. 5.4.3. Ironing is effected by the radial compressive stress acting on the material deriving from the tensile stress exerted by the punch; as a cold-worked thickness reduction of more than 50% is effected, copious amounts of a good lubricant are needed, additionally to limit tool surface damage and provide a good can surface. Tool manufacture must be to a very high standard, and carbide inserts are generally used to give long tool life. The substantial thickness reduction involved causes a corresponding substantial increase in body height, but with an irregular rim. Excess height is usually trimmed off in rotary trimmers. Additional lubricant is applied at each of the three ironing stages. Kohn[(16)] discusses the stringent demands made on tool design and manufacture, due to the thinness of tinplate,

and the especial economic need for very long tool life. Tool surfaces need to be very smooth, and blending radii very accurately formed. Examples of tool design are given in his paper. Punches and dies are usually retained as matched sets to maintain maximum cup accuracy. The process is considered in detail by Panknin.[21]

5. Before lacquering and decorating can be carried out, the heavy residual lubricant which will contain small particles of swarf must be very effectively removed; this can be carried out in hooded mat-conveyor type cleaning units employing water-based detergents or organic solvents (which are liberally sprayed inside and outside the cans) to ensure near absolute cleanliness, the final stage requires very thorough washing — generally with a final wash in deionised water — and quick drying, using hot air. If needed, a chemical surface treatment facility (chromate, phosphate type) can be incorporated as a final stage.
6. When required to be externally decorated, a base coat is first applied and oven dried, then the printed design, using up to four colours, by the rotary technique described in section 6.4; finally, a varnish coat is applied, followed by stoving. To avoid damage to the print before it is dried, the cans are usually transported through the oven by means of "peg" conveyors.
7. The can rim is then flanged outwards to enable the top end to be seamed on after filling by the packer; flanging is almost always coupled with necking (Fig. 5.4.4) which reduces the overall diameter across the seamed end to below that of the can body wall. This is accomplished by die-necking and spin-flanging, or combined roll flanging and necking-in (Fig. 5.4.5). By this means, blows from the double seam on to adjacent can walls during transport can be avoided, thus reducing the danger of damage. It also provides a significant saving in endplate material resulting from the smaller diameter end needed, and also allows more effective packing methods to be adopted.
8. Finally, the internal surface is spray lacquered (as described in section 6.4) and then stoved; for soft drinks two coats are usually applied, employing precise control of film weight and spray distribution to ensure absolute minimum metal exposure. For some of the less aggressive beers, one coat may be considered adequate.

The primary cupping press will operate at up to 200 strokes per minute, but to achieve outputs approaching 1000 cans per minute requires very good quality plate free from significant defects as discussed later, in addition to high-quality tooling and expert setting. Elaborate instrumentation is fitted to these cupping presses to ensure instant stoppage on case of any malfunction.

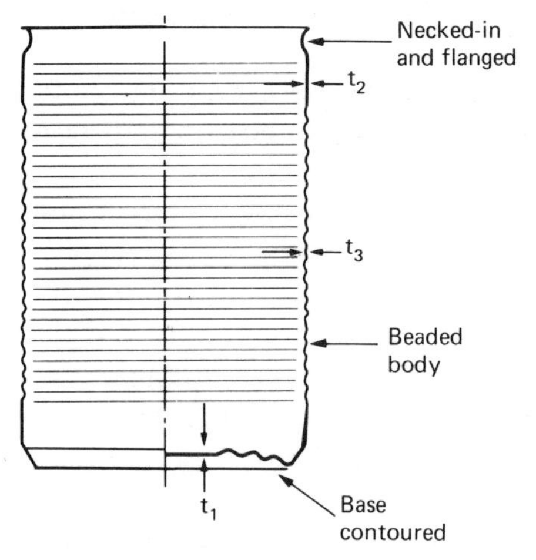

FIGURE 5.4.4 CAN BODY BEADED AND BASE CONTOURED FOR PROCESSED FOOD CAN

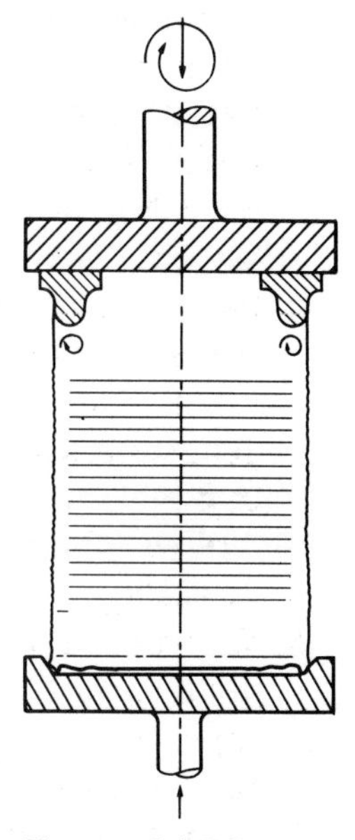

FIGURE 5.4.5 SPIN-FLANGING OF DWI CAN

Precise location of the blank and of the first stage cup with the subsequent tooling is essential to avoid non-symmetrical cups, and thus excessively variable wall thickness.

Thickening of the cup wall towards its rim, which normally occurs during drawing, is reduced in extent by the process of ironing, but a small increase in thickness from near the base upwards to the rim will persist (about 0.03 mm).

By appropriate tooling design the wall portion adjacent to the can rim will be slightly thicker to assist necking-in and flanging without metal failure ("split" flanges). As the base thickness will remain unchanged and not cold-worked, its resistance to high internal pressure will not be adequate, so a doming tool is fitted to the base of the punch which, with a corresponding base tool, will form an inward dome to a desired contour. It is vital that the base does not dome outwards during filling and pasteurising, to avoid troublesome rocking of the can at high processing speeds. The technology is thus eminently suitable for cans to be subjected to high internal pressure.

The ironing dies, usually made of carbide, are supported by steel housings to avoid being subjected to dangerous tensile hoop stresses. Campion[38] states that normally each ironing ring is so located in relation to the next that the body will have entered the second before fully leaving the previous one, thereby avoiding a dangerous sudden release of strain energy which would otherwise cause it to approach the next die at far too high an impact velocity. Smooth passage through the tool sets ensures more accurate components. Kohn[16] reports that tinplate is too thin to be ironed by more than one ring at a time.

Although copious volumes of lubricant are used at each ironing stage, and holes are drilled radially in the spacing rings to assist flow of coolant, metal temperature rise can be appreciable on a macro scale — up to 70°C is not uncommon — and therefore much higher on a micro scale. It has been reported that this rise in temperature brings about a significant improvement in ductility, thus easing ironing loads.

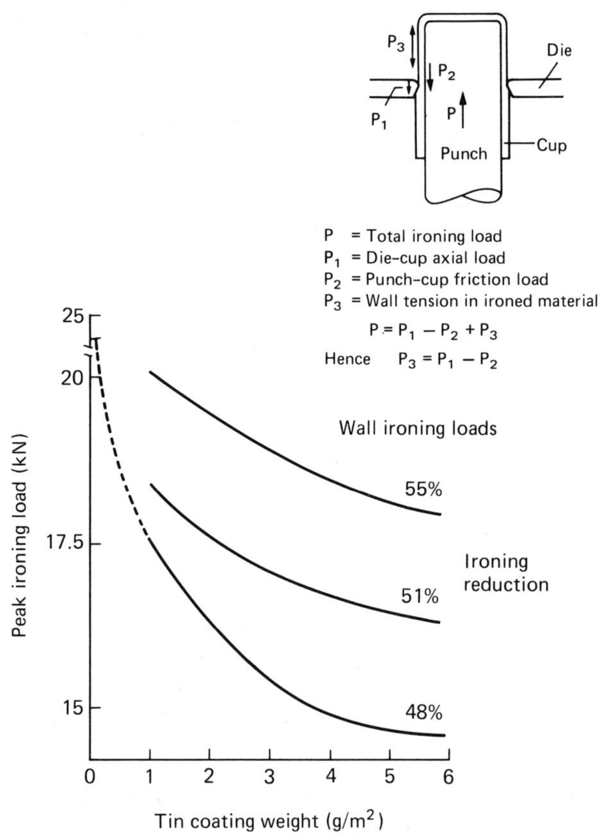

FIGURE 5.4.6 RELATIONSHIP BETWEEN PEAK IRONING LOAD AND TIN COATING WEIGHT

It has been widely reported that the tincoating confers a significant benefit on the wall ironing operation; e.g. Duckett *et al*[17] and Duckett and Thwaites.[18] Fidler[19] quotes the relationship found in his work between decrease in ironing peak loads with increasing tincoating weight up to 6 g/m^2, for three levels of ironing reductions, and concludes that the ironing load required for untreated blackplate similarly lubricated is substantially higher (Fig. 5.4.6).

As would be expected, the external tincoating is substantially damaged by wall ironing, even though it appears visually very lustrous and smooth. Scanning electron microscopic examination shows tin distribution to be very patchy, and when removed the underlying steel surface is found to be deeply scored in the ironing direction, the grooves appearing to be filled with tin. Extent of damage increases with each ironing stage; the internal surface is comparatively much less damaged, as would be expected from the mode of metal movement involved; it increases up to virtually 100% iron exposed towards the can rim. These effects are shown in Figs. 5.4.7 and 5.4.8.

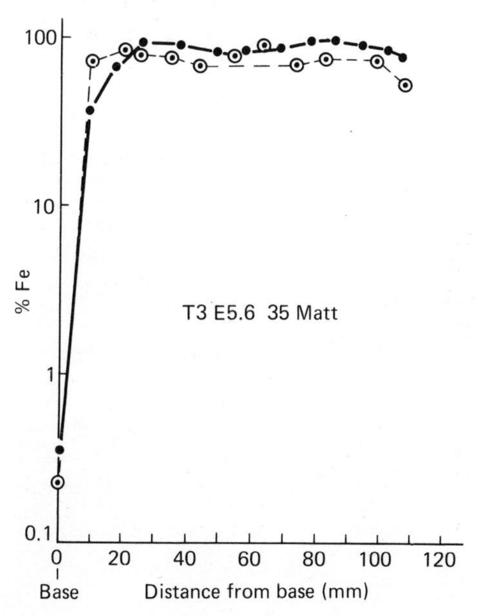

FIGURE 5.4.7 IRON EXPOSURE ON TINPLATE DWI CANS

Thus the residual tincoating cannot be expected to offer much protection, and general experience has shown that the products more aggressive to iron are likely to cause severe pitting or perforation without extra protection than in three-piece cans. Thus in general two internal lacquer coats need to be applied very evenly

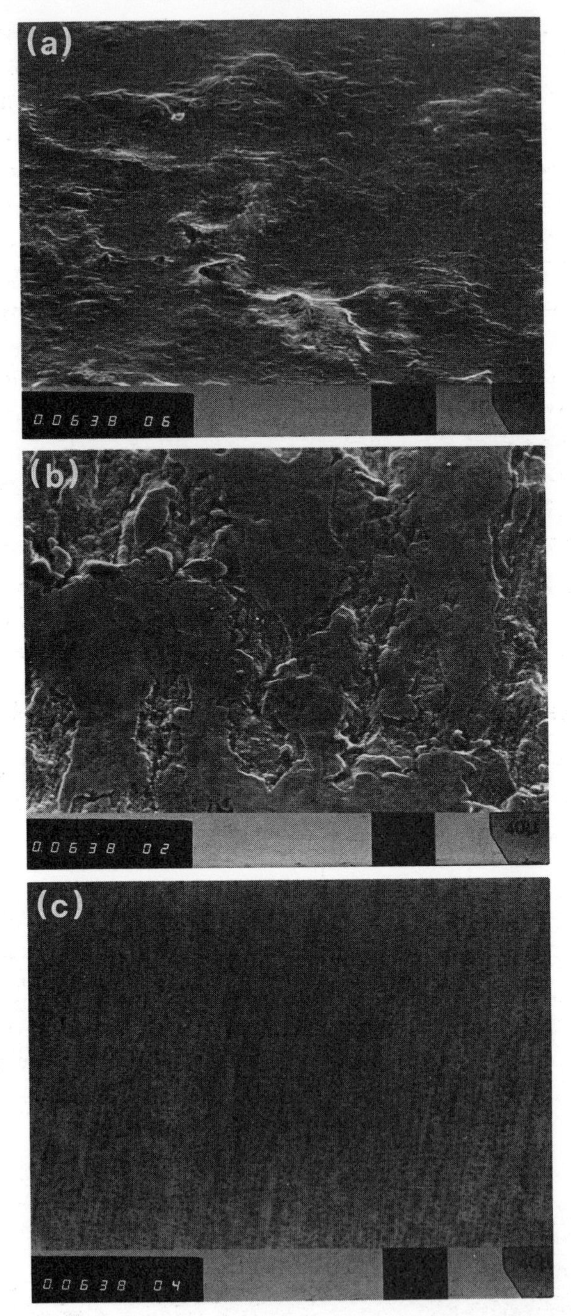

FIGURE 5.4.8 (a), (b) and (c) ILLUSTRATE EFFECTS OF SUCCESSIVE STAGES OF IRONING ON TIN DISTRIBUTION

over the whole surface; this is judged by enamel rater (metal exposure) values. Given this treatment, most corrosive soft drinks are also fully satisfactory, as has been confirmed by considerable commercial experience. Some of the more inert soft drinks may be satisfactory with a single sprayed lacquer coat. Some beers are being packed in cans given a single sprayed lacquer coat.

Packing processed foods in DWI cans imposes two stringent conditions; lacquer fracture must remain very low after the heat sterilising process, and it must be compatible with the food.

The difficulty is even greater when the food is of the sulphur staining type, as the protective passivation film formed after electrotinning will have been almost completely destroyed; the lacquer must additionally be capable of being fully stoved within 1 minute, in comparison with not less than 10 minutes when stoving roller coated lacquer. Nonetheless, good practical performance is being achieved in some areas.

Carter *et al*,[20] in their study of the effects of tincoating weight in the range 1.1g to 22.4 g/m^2 on the porosity and tin distribution in DWI cans, found similar effects, the heaviest tincoating showing the least damage. Chemically treated cans after DWI, made from plate with extremely light tincoatings, showed a better performance in tests than the untreated cans carrying the heaviest tincoating.

5.5 DRAW AND REDRAW (DRD) PROCESS

The principle of this method is that a "first operation" drawn cup is redrawn to one of smaller diameter, and thus of greater height; this may be achieved in one or two redraw operations, as shown in Fig. 5.5.1.

FIGURE 5.5.1 DEEP DRAWN CANS SHOWING SINGLE REDRAW AND TWO REDRAW SEQUENCES

The advantages over the traditional three-piece can, as with a DWI type, are that a container free of a side seam and with only one double seam is produced. It has the particular advantage over DWI in that prelacquered plate can be used, thus avoiding costly and difficult post-cleaning and spray-lacquering; it is usually lacquered on both surfaces. An internal lubricant is added to the lacquer on roller-coating; additional drawing lubricant is precisely applied prior to drawing.

The greatest can height:diameter ratio that can be produced by multiple drawing is usually less that than by DWI; the DRD can is favoured for the shorter food cans. As a deliberate change in gauge does not occur in this process, the base material will usually be strengthened to the level required by coining a profile into it.

Investment cost is much lower than that needed for DWI line of similar output, being comparable to that for three-piece can manufacture. The line is compact and short, and is yet a high speed unit, producing up to 1000 cans per minute to precise dimensions. Although the practice is to use scrolled plate, and to aim for minimum shred width between the blanks, plate saving is significantly less than that provided by the DWI method. DRD is being increasingly used for the manufacture of shorter processed food cans, employing either two-stage (one redraw) or three-stage (two redraw) operations as capital and operating costs are more favourable in these cases. The main stages of the DRD process can be summarised as follows.

A typical example of a DRD produced can is one having a diameter of 73 mm, and finished height of 110 mm, redrawn in two stages. As the plate thickness will remain sensibly constant, the area of the circular blank required will be approximately equal to that of the untrimmed finished container; it will vary according to the effects of the tool radii and the particular material properties. Final adjustment of blank cut-edge diameter will be made during development of experimental tooling and material testing. The optimum plate thickness will depend on can size and application required; it is usually within the range 0.20-0.22 mm. As a flat base at this thickness will not have adequate resistance to internal pressure, it will be profiled to an optimum contour at the end of the last drawing stroke, as in the DWI process, by attaching appropriate tooling to the base of the punch and die. Care must be taken to ensure that the coining tool radii are not too sharp, as this can lead to lacquer fracture. Separate presses are used for each stage, finally trimming the irregular flange.

Panknin[21] has described in detail the theoretical principles of drawing and redrawing (and also those of wall-ironing); it includes consideration of the major factors involved, covering

tool geometry in terms of material thickness; its surface finish and that of the plate,
calculation of blank diameter for a given can size, and of blank-holding force.
limiting drawing ratio (blank diameter: punch diameter), in terms of blank diameter, punch diameter, and material thickness.

The tallest can height possible is dictated by the limiting drawing ratio. It is usual to work to as near the maximum as practical, taking due account of material variation, as this will reduce the number of redrawing stages. The LDR value varies according to many tooling and plate parameters, and to the plate lacquering system. The highest LDR value is obtained with a punch profile radius, R_p, of about $0.1D_p$ and, particularly for thin tinplate, the optimum die radius R_d of at least $8t$ (t, the plate thickness); a larger radius increases the susceptibility to flange wrinkling, while a sharper radius will stretch the material beyond the drawing die radius, thus increasing the danger of lacquer damage. The r value of the material, although having substantial effects on tooling performance generally (these are described later), has only a small effect on the maximum draw, LDR. Accurate tool setting is also very important.

Lubrication conditions, coupled with tincoating weight and lacquer type, have marked effects of LRD values; the increase achievable by improved lubrication, higher tincoating weight and internal lubricant in the lacquer coat have been investigated in detail by Siewert.[14] He concludes that to achieve adequate freedom from lacquer fracture, an internal lubricant must be incorporated into the lacquer on roller coating. Careful choice of lacquer type is also essential, as it must be related to the type of passivation treatment used after electrotinning to ensure maximum lacquer adhesion, including possible deterioration after heat processing of the filled can.

The width of the drawing gap (the clearance between punch and die) ($R_d - R_p$) can be critically important when the draw ratio is near its maximum, as under this condition greatest thickening will occur towards the edge of the flange; if this were to be greater than the clearance, considerable lacquer damage would result.

Siewert's work[14] also indicates that plate surface should not be very smooth, to assist in maintaining a continuous lubricant film during drawing through the dies, but it is vital that tool surface should be as smooth as possible.

The alternative reverse redrawing technique, illustrated in Fig. 5.4.2 (c) in which the cup is turned "inside-out", is mechanically more efficient, as it contains less bending/unbending (less "redundant" work) in the metal flow, but it is not used for DRD as its effects on lacquer damage would be appreciably greater.

Most of the comprehensive papers on the DWI technique, referred to earlier in that section, and included in the BSC literature review[24], also contain valuable discussion of aspects of the DRD process. Particular attention can be drawn to that by Abe *et al*[22] and by MacPhee,[23] who derived a practical method for determining the limiting parameter of the redrawing process for cans drawn from thin blanks of low carbon steel; it presents a method for predicting the number of draws required to form a particular can size. The BSC survey[24] contains many other valuable papers on these processes, e.g. by Campion.[38]

5.5.1 The Metal Box DRD process

As the output from large crank presses is limited, due to the heavy reciprocating masses inherent in their design, Metal Box plc decided to develop redraw close-coupled presses based on rotary continuous operation.[19]

The lacquered scrolled plate is fed directly into a standard cupping press, usually double-action type, in which the first stage cup is drawn; these are then fed directly into an eight-head rotary press in which the cup is redrawn into one of smaller diameter and greater height. If one redraw stage only is required for the particular can size, the base is contoured by coining at the end of the punch stroke. The irregular flange is then trimmed in a second, similar rotary press to the width specified for later double seaming the top end. When two redraw stages are required for taller cans, the integrated line will usually contain three rotary presses; alternatively, flange trimming can be carried out as a final separate operation, using conventional high-speed trimmers. Line speed is only limited by that of the initial cupping press, currently about 700 cans per minute. The rotary presses are capable of operating at significantly faster speeds.

DRD is sometimes referred to as a clean process, as it does not involve removal of excess lubricant by washing, degreasing and drying. It does, however, require accurate control of the amount of drawing lubricant, for which a precise roller applicator has been developed. A line of this type is very compact, occupying little space in comparison with a standard three-piece can line, or a DWI line.

5.5.2 Effects of plate characteristics in the DWI and DRD processes

Several plate characteristics have a marked effect on its behaviour in Drawing and Redrawing, and in a similar manner on Drawing and Wall-ironing, but to different extents. The physical and mechanical requirements and their metallurgical control have been described in detail by Jenkins.[25] As both processes are high speed, their economics are vitally dependent on high production efficiency and low scrap levels; absence of defects and optimum properties are paramount in minimising down-time.

Plate effects can be summarised as follows:

	DWI	DRD
1. *Steel cleanness.* A low level of non-metallic inclusions (NMI) is vital, as they can become sites of stress concentration during heavy deformation, and can lead to metal fracture. Two significant failures of this type are microfractures in drawn cup walls and flange cracking. A low incidence is generally assured by specifying the more expensive killed steel (AK), preferably the continuously cast type. Methods of evaluation cleanness and appropriate standards are described by Jenkins and others.[26-29] Their incidence and other methods of evaluation have also been discussed by many other workers.[30-32] Use of rimmed-capped (RC) steels has also been studied, but undoubtedly a high standard of steelmaking is absolutely vital in that case, if significant scrap is to be avoided. Difference in NMI levels due to method of casting are illustrated in Fig. 5.5.2.	Vital	Vital
2. *Steel chemistry.* This is not normally specified by the canmaker, but the		

Cont.	DWI	DRD
steelmakers control it to the narrow limits described in Chapter 2, to ensure that the appropriate temper grade can be achieved.		

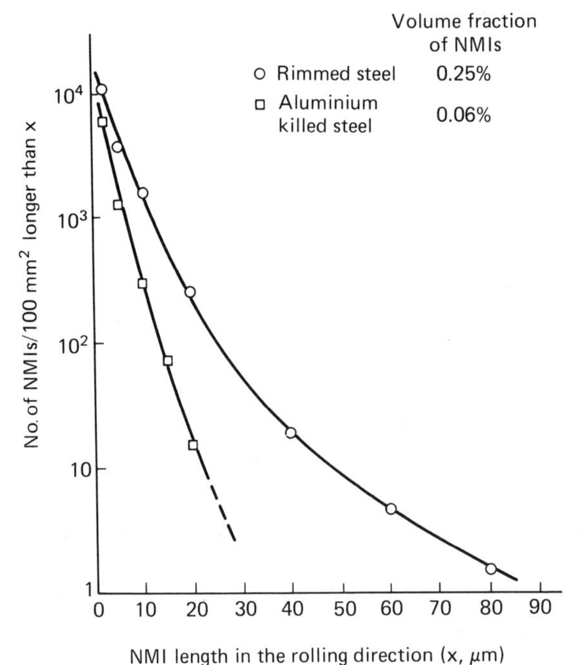

FIGURE 5.5.2 FREQUENCY/SIZE DISTRIBUTION OF NMIs IN STEELS PRODUCED BY THE BOS STEELMAKING PROCESS

3. *Drawability and minimum earing characteristics.* These must also be be satisfactory in many respects. The cold rolling and annealing conditions have an important bearing on earing tendency (reflected by Δr value), r value, and ferrite grain size, as discussed in Chapter 2. Excessive earing, (Fig. 5.5.3) reduces effective can height; it can cause objectionable metal slivers due to pinching off the ear ends; greater wrinkling in the thinner ear zones; danger of buckling and folding near the

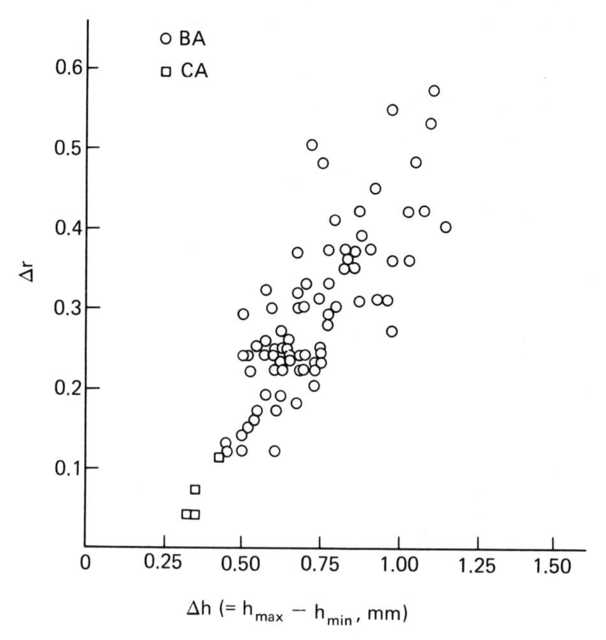

FIGURE 5.5.3 RELATIONSHIP BETWEEN Δr AND EAR HEIGHT (Δh) FOR 0.22mm RC-BA AND RC-CA TINPLATE

cont.	DWI	DRD
(thicker) ear troughs; and a greater danger of split flanges.		
Higher r_m values result in several benefits — greater resistance to thinning in the cup zones subjected to higher tensile stress, particularly the cup wall and over the punch nose, compressive stress on to the draw ring to avoid wrinkling would be lower, and thus a reduced tendency for wrinkling. These lead to a higher β_{max} value and a taller cup. Typical values for various plate types, reported by Jenkins are:	Desirable	Vital

Product	r_m value
Aluminium killed steel, batch annealed (AK-BA)	1.0-1.8
Rimmed steel, batch annealed (RC-BA)	0.8-1.1
Rimmed steel, continuously	

cont.		DWI	DRD
annealed (RC-CA)	0.7-1.0		
Aluminium killed steel, continuously annealed (AK-CA)	0.7-1.0		
Double reduced (DR) tinmill products			
Too large a ferrite grain can give an "orange-peel" surface finish on severe drawing, which could seriously interfere with lacquer continuity, and be aesthetically unacceptable. These effects have been described in several papers.[21,33] The excessive ferrite grain sizes only normally occur when processing the AK-BA product to provide high r_m values; this relationship raises the possibility of independent means of grain refinement, to allow material of high r_m values coupled with smaller grain size to be produced.		Not essential	Vital
4. *Thickness variation.* Plate gauge has a significant effect on cup height, and the tolerance required is invariably tighter than that for "run of the mill" forming operations; transverse guage variation (between the "crown" and "feather edge") is also usually specified to a lower tolerance than usual.		± 5% (or less)	± 7½%
5. *Pinholes and welds.* The presence of these faults could lead to tearing of plate in the cupping operations, leading to possible serious tool damage and serious "down-time" due also to plate tearing in the DWI operation. As prelacquered scrolled plate is generally used in the DRD process, these faults should have been completely eliminated during earlier coil processing.		As complete elimination as possible	

cont.	DWI	DRD
6. *Surface finish.* This is not critical in the DWI process, as canmakers' specifications range from "smooth finish" ($R_a \simeq 0.3\ \mu m$) up to "80^{m}" finish ($R_a \simeq 2.5\ \mu m$). A slightly rough grade ("stone" finish) is favoured usually for the DRD operation, as this is said to assist lubrication performance.		
7. *Tincoating weight.* The optimum weight in the case of DWI, for which the plate is not lacquered, is primarily determined by the wall-ironing conditions and lubrication system used; a widely used specification is 2.8/2.8, or the nonstandard grade 2.2/3.4. For DRD the tincoating grade is usually 5.6/5.6 or 2.8/2.8, primarily according to the product, but the passivation treatment (and lacquer system) is also important. Chrome/chrome oxide ("tinfree steel") coated plate is also used.		
8. *Passivation treatment.* As mentioned earlier, this can be critical in the DRD method; the type generally found to give maximum lacquer continuity, coupled with specially developed lacquer systems, is one based on dip treatment (non-electrolytic) in a sodium dichromate solution.		
9. *Coil geometry.* As the residual shreds after cutting out the blanks in the cupping press are controlled to a very narrow width, coil quality must be better than average, being reasonably free from the faults (especially edge wave) mentioned in Chapter 3.		

Plate meeting these optimum characteristics will contribute substantially to efficient operation of each of these high-speed critical operations, and many can offer the opportunity for reduction in gauge, blank size, etc.

Many changes have been made to the specifications originally adopted, although the need for a good clean steel still applies; wherever supplies are adequate, continuous (strand) casting is usually a first choice, because of the better casting yield it provides, and, some maintain, overall it has the edge over ingot cast material in respect of cleanness. This is dependent on many factors, especially secondary steelmaking techniques, and on general expertise.

In other respects there has been a gradual change to higher temper grades, and to continuous annealing; e.g. CA T61 ("T4") is already used commercially in the DWI technique, while some DRD operators are using double-reduced plate DR 550/620 ("DR8/DR9") BA and also DR620 CA ("DR9").

Lower tincoating weights are also being used in comparison with the earlier grades, in both processes, and also chromium/chromium oxide coated steel sheet in DRD.

Commercial viability of any new plate specification will be very dependent on the equipment used, lubrication system, lacquer type, etc.; all of these are closely interrelated and any change needs thorough testing in the process; and of packed containers. Trials are often carried out in conjunction with major customers.

A view of possible future developments is taken in Chapter 9.

5.6 "EASY-OPEN" (EO) ENDS

These ends are broadly of two types: (i) those providing a pouring aperture, roughly pear-shaped, for dispensing liquid products such as beverages, fruit juices, oils, etc., and (ii) those giving a near full aperture opening for removing more solid products (thick soups, fish in oil or sauce, meats, nuts, etc.).

Although a few types of easy-opening devices have been available for many years, e.g. the key opening scored strip found in solid meat or shallow fish cans, housewives have consistently complained of the difficulty in opening all food cans, particularly when the opening tool had become far too blunt over many years of use.

The first modern type of easy-open end to be developed was that formed in aluminium ends for beverages; it has been claimed that this development was a major factor in the considerable growth in use in recent years of the beer and soft drink can. This design, generally referred to as a ring-pull end, consists of a scored portion coined into the end panel and a levering tab partly formed in a separate progression and then rivetted into one end of the scored area. Opening is readily effected in two stages, by (i) lifting the tab, causing the residual metal at the score adjacent to the tab to shear, and then (ii) by gripping the ring and tearing back along the whole score length.

The aluminium alloy used (Type 5182 H19) for the end has a gauge in the range 0.30-0.35 mm (0.012-0.014 in); as this has to be scored to a depth about half its original thickness to provide a reasonable tearing load and yet have acceptable resistance to fracture under high internal pressure and shock loading, a lower thickness tolerance than normal is specified, together with high precision toolmaking. The indent is designed so that the residual edge after tearing open does not constitute a real hazard of cut fingers.

Many attempts have been made over the years to design a pouring aperture type made from tinplate; this had not been completely successful until recently, as (i) the score depth needed to give reasonable tearing forces is substantially greater than that for aluminium, up to two-thirds of the plate thickness, and (ii) difficulty in keeping tearing to the scoreline, especially around sharp corners of the pear shape — this being due to the plate rolling direction and directional NMIs. This development is still proceeding; in fact a 202 (54 mm) diameter end made in tinplate has been available commercially over the last few years for cans packed with hot-filled drinks; larger diameter ends are being developed. As described below, the full-aperture type tinplate ends have been in commercial use for longer periods.

A second type of pouring easy-open end has been developed by the Australian Iron and Steel Co., termed the "Presto". It has been described in detail by Wiltshire.[(34)] As Fig. 5.6.1. shows, the design does not incorporate a separate lever; it consists of two circular tabs, one smaller than the other, formed into the end panel by scoring almost completely around the aperture circumference, leaving just sufficient unscored portion to function as a hinge. After fabricating the end, a sealant is applied to the scored cut edges internally, and a lacquer to the external cut edges to minimise corrosion. A standard end is used for conversion, and fewer press operations are needed than in the ring-pull tear type. Aluminium alloy material was first used, but later Presto ends made from tinplate have also been developed. The internal sealant compound was developed from a conventional plastisol to satisfy the criteria of a pasteurised beverage container.

FIGURE 5.6.1 AN EASY-OPEN "PRESTO" END

The full aperture type EO end is similar in basic design to the ring-pull pouring type, in that an integrated ring tab is used first to pierce the score and then to tear along the score line; press fabrication is also similar. The score line is usually positioned as close to the inside double-seam wall as possible, so that the residual lip width is minimal and the diameter of the opening the maximum. Close control of scoring conditions is vital to ensure adequate resistance to bursting and yet without needing an unduly high tearing load to open. Residual thickness at the score line will also be dependent on plate thickness and it is customary to specify a lower than usual tolerance on its variability, e.g. ± 7½%. Kohn[16] states that residual thickness must be accurate to 5% of nominal, and as tool parts cannot be made to this accuracy, stop blocks are incorporated as shown in Fig. 5.6.2; a setting allowance is included for plate thickness variation and for the elastic deformation of the tooling system.

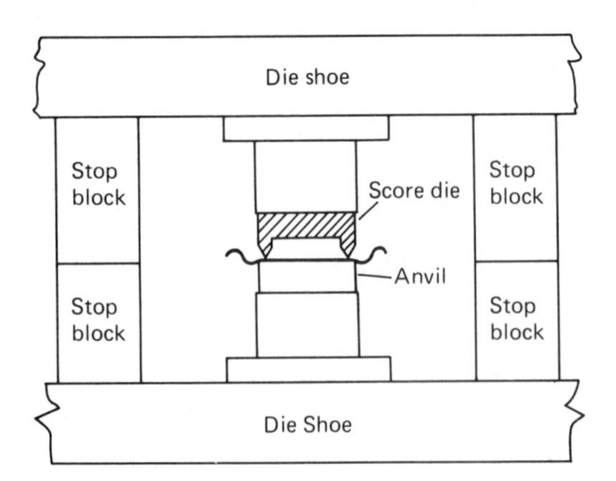

(a) stop blocks introduced to ensure correct gap settings

FIGURE 5.6.2

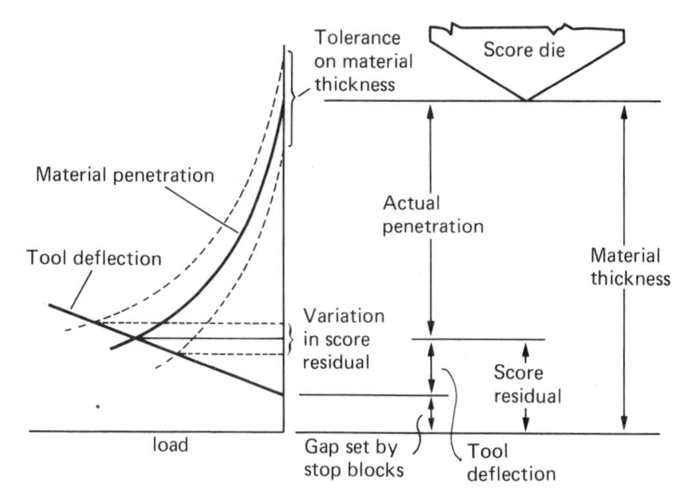

(b) gap measured without material present allowing the tooling to compress when it penetrates material

FIGURE 5.6.2

FIGURE 5.6.3 A SELECTION OF EASY-OPEN ENDS

A high standard of steel cleanness is also required, especially when the EO end is required for processed food cans. The score indent can also have an effect on ease of opening and on sharpness of residual lip; its form will vary between manufacturers, some preferring a fairly sharp but radiussed V-notch, whilst others use one having a positive land radiussed into the conical section. Particular attention must also be paid to metal exposure resulting from internal lacquer fracture at the score if the ends are to be used for food cans; sometimes repair is effected by post-lacquer spraying the end after conversion.

These EO ends are commercially available in circular and rectangular form for a wide range of dry and processed food products. A particular application is that for food packs dispensed warm from institutional vending machines. A selection is shown in Fig. 5.6.3. Similar ends made from aluminium are also widely used, especially for the flatter drawn cans for fish products.

5.7 PRESSES AND TOOLING

The presses used for metal-forming processes can be divided broadly into two types: mechanically and hydraulically operated. For high volume press-forming the mechanical type is usually preferred, as this is fast in operation, but with a comparatively shorter stroke than is possible with the hydraulic type; the latter is capable of providing long stroke operation but at significantly slower rates.

Mechanical presses can be further subdivided into several types, according to their design principle:

(i) crank and eccentric,
(ii) toggle,
(iii) screw.

Each of these has particular advantages and disadvantages; detailed descriptions and evaluation of their characteristics can be found in many specialist books on presses, such as Schuler's *Metal Forming Manual*,[35] Grainger's *Presswork and Presses*[36] and *Techniques of Pressworking Sheet Metal*.[37] Other systems of classifying presses are also used, according to their strength and capacity; single, double and triple action, etc. The terminology used for their component parts will be found to vary.

Crank-type presses are widely used in high-speed press work. Energy is supplied by means of an electric motor through a massive flywheel to a crankshaft. As the stroking speed can be very high — several hundred strokes per minute — it is vital that the flywheel does not slow down excessively on each stroke operation (not more than 15-20% loss is usual), so that the desired operating speed is safely maintained without danger of stalling and overloading the electric motor. Rotary motion of the flywheel is converted through a crankshaft into the reciprocating motion of the press slide by means of the crank. The Schuler book describes these aspects and gives formulae for calculating speed of the slide and its position in relation to that of the crankshaft, the force exerted at various positions and the work done, permissible reduction in flywheel speed, and motor capacity required for various cases. Selection of a press type for particular operations and overload precautions are also considered. A double action press will have two slides, one for the blankholder and the other for the drawing operation, and these will move independently of each other.

The press frame can be of many forms; Fig. 5.7.1 illustrates two basic types, a gap (or C) frame, and a double-sided press. As with any metal and structure under stress, a change in shape will occur when the press is under load, to a degree dependent on its rigidity. This deflection or "spring" can have several unfavourable effects, such as excessive wear of cutting edges, resulting in more frequent tool changes than necessary, and quality of components will suffer; these effects are shown schematically in Figs. 5.7.2 and 5.7.3. Kohn[16] emphasises that lateral stability is essential to preserve precise alignment between individual units in the case of multiple

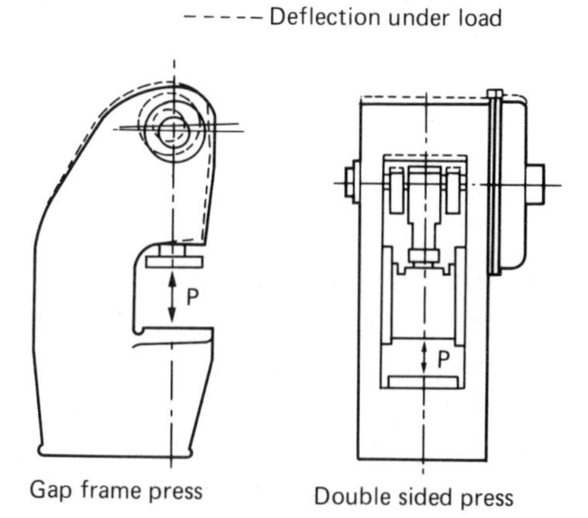

FIGURE 5.7.1 SPRINGING CHARACTERISTICS OF DIFFERENT PRESS FRAMES

tools, and also extreme rigidity is needed to minimise elastic distortion in the direction of operation; this is achieved by four-post construction and a solidity of scantings equal to that in higher duty presses used for thicker material. In view of the very demanding requirements of these processes, highly sophisticated presses have been developed by many major canmakers in conjunction with specialist press manufacturers.

Most books on presses also describe the principles and characteristics of hydraulic presses.

Techniques of Pressworking Sheet Metal also contains a chapter on mechanical handling devices.

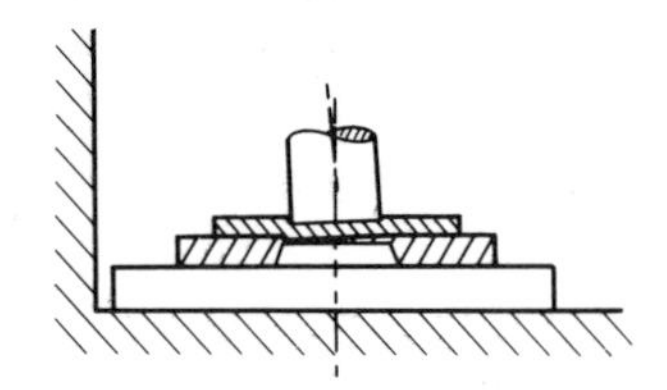

FIGURE 5.7.2 MISALIGNMENT OF PUNCH CAUSED BY DEFLECTION OF GAP FRAME PRESS

Tools are usually of the double action type; a DA design and also that of single action tooling are illustrated in Fig. 5.7.3.

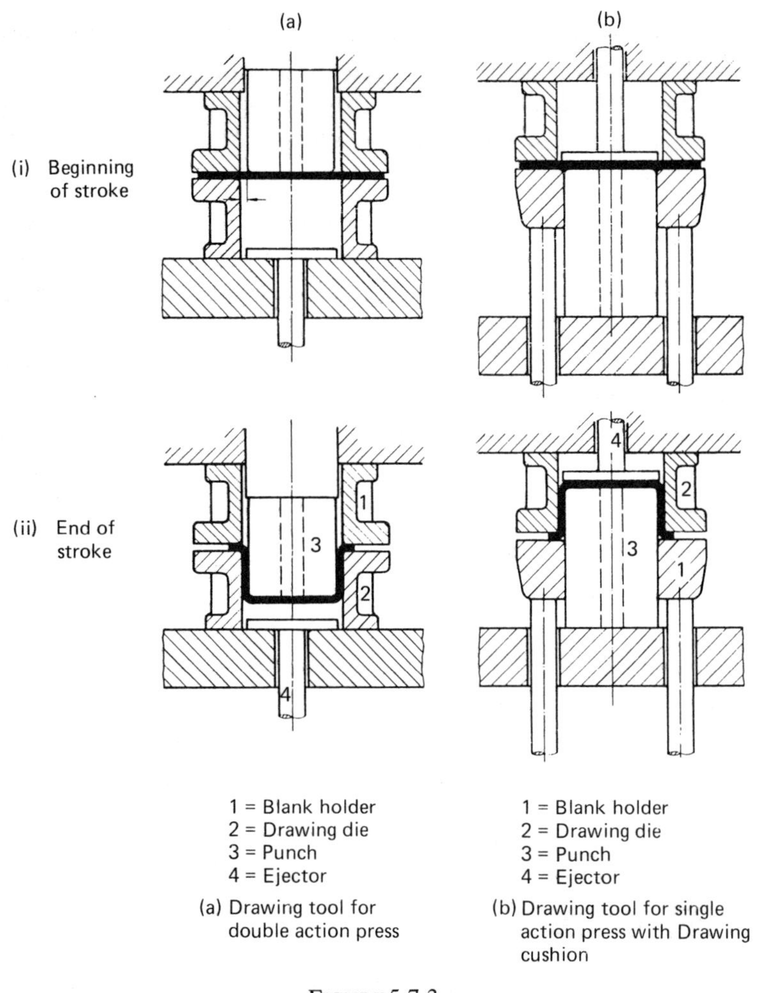

FIGURE 5.7.3

Both the design and the standard of manufacture of the tooling used in DRD and DWI processes are of paramount importance. As each is a high volume production technology, their economic viability is critically dependent on efficient production with minimum down-time and scrap levels. Kohn[16] considers these aspects in some detail.

The demand for optimum tool design is especially critical, as

(i) the material thinness and comparatively high H/D container ratios require the cans to be drawn at typically around 93% of the maximum LDR;
(ii) degree of redundant (wasted) work must be minimal;
(iii) to minimise the danger of lacquer damage, radii must be blended smoothly into flats, and surfaces must be ultra smooth;
(iv) long tool life (low unit cost) by employing means for limiting tool wear;
(v) minimum tool stocks, compounded by multiple tooling in cupping presses; interchangeability and rapid tool changing.

These call for the development of particular expertise in design, with a corresponding standard of tooling manufacturing accuracy.

The fine tolerances have required specially trained personnel, and the highest quality machine tools installed in precision toolrooms having almost clinical atmosphere, free of dust, with close control of temperature and humidity, and development of sophisticated equipment for fine measurement.

Many of these important aspects are also discussed in the Schuler[35] handbook. Experience has shown that close attention must be paid to these factors to ensure economic viability and high-quality containers.

Blazynski[42] provides an analytical approach to the fundamental principles of tool profiles in relation to metal flow.

A British Steel Corporation review of literature[24] includes references also to papers on tooling aspects.

The need for use of good steels in the manufacture of specialised tooling is emphasised in most of these books on presses and tooling. But as this is a very specialised subject, special steel manufacturers are generally very happy to offer advice on most appropriate steel grades for various applications.

REFERENCES

1. R. PEARCE, The Tensile Test as a Guide to Forming Properties, *Sheet Metal Ind.*, **40,** (433) (1963) 317.
2. J.O. WARWICK and J.M. ALEXANDER, Prediction of the Limiting Drawing Ratio from the Stress-Strain Curve, *J. Inst. Met.*, **91,** (1) (1962) 1.

3. D. HULL, *Introduction to Dislocations,* 2nd Ed. Pergamon Press.
4. GEO. E. DIETER, *Mechanical Metallurgy,* 2nd Ed. McGraw-Hill, New York, 1976.
5. G.A. JENKINS, Demands Two-piece Can Technology Place on the Properties of Tinplate. First International Tinplate Conference, ITRI, 1976.
6. H. HU, *Met. Trans. A.,* 6A, 1975, 945, 2307.
7. J. WILLIS, *Deep Drawing* (a summary of Prof. H.W. Swift's studies *Proc. Inst. Mech. Engrs.* **165,** 1951, 199) Butterworth, London, 1954.
8. J.M. ALEXANDER, *Metal Reviews,* **5,** (1960) 349.
9. J.L. DUNCAN and W. JOHNSON, Approximate Analyses of Loads in Axi-symmetric Deepdrawing, Proc. 9th International MTDR Conference. Pergamon Press, 1969.
10. T.Z. BLAZYNSKI, *Metal Forming, Tool Profiles and Flow,* McMillan, London, 1976.
11. W. JOHNSON and P.B. MELLER, *Engineering Plasticity,* Van Nostraud Reinhold, London, 1973.
12. H.W. SWIFT, *J. Inst. Met. (London), 1952.*
13. G. SACHS and K.R. VAN HORN, *A.S.M.* 430, 1940.
14. J. SIEWERT and M. SODEIK, Seamless Food Cans made of Tinplate, First International Tinplate Conference, ITRI, London, 1976.
15. L. GARY, Selecting a Mechanical or Hydraulic Press for Your Particular Job. Bulletin 50-B, E.W. Bliss Co., Hastings, Michigan, USA.
16. E.O. KOHN, Tooling Aspects of High-Speed Can Production, *Sheet Metal Ind.* June 1976, 331.
17. R. DUCKETT, B.T.K. BERRY and D.A. ROBBINS, Effect of Tincoatings on the Drawability of Steel Sheet, *Steel Metal Ind.*, Sept 1968, 666.
18. R. DUCKETT and C.J. THWAITES, A Note on the Wall-ironing of Cup Drawn from Tinplate, *Sheet Metal Ind.*, April 1971, 274.
19. F. FIDLER, Two-piece Container Developments — Some Influences of Tin, First International Tinplate Conference, ITRI, London, 1976.
20. P.R. CARTER, L.L. LEWIS and M.V. MURRAY, Surface Properties of Drawn and Ironed Tinplate Containers, First International Tinplate Conference, ITRI, 1976.
21. W. PANKNIN, Principles of Drawing and Wall-ironing for the Manufacture of Two-piece Tinplate Cans, First International Tinplate Conference, ITRI, London, 1976.
22, H. ABE, Trans. BSC SMD 680, *Kawasaki Steel Tech. Rpt.,* **8,** (1) 1976.
23. J. MACPHEE, *Sheet Met. Ind.* **53,** (11) 1976, 427; **54,** (1) 1977, 75.
24. A.H. PHILLIPS and R.L. DAVIES, A Survey of the Literature Concerning the Manufacture of Two-piece Cans made from Tin Mill Products, by the DWI and DRD Routes, British Steel Corporation, Port Talbot, 1979.
25. G.A. JENKINS, G. JEFFORD, and D.W. EVANS, Demands Two-piece Can Technologies Placed on the Properties of Tinplate, First International Tinplate Conference, ITRI, London, 1976.

26. R. KIESLING, *J. Metals,* Oct 1969, 48.
27. P.A. THORTON, *J. Mat. Sci.* **6,** 1971, 347.
28. 75th BISRA Steelmaking Conference, Scarborough, 1970.
29. BSC/BISPA Conference, Scarborough, 1973.
30. N.E. MOORE, A. O'CONNOR and G.L. THOMAS, *J. Iron & Steel Inst.,* **118,** 1971, 600.
31. N.E. MOORE and G.L. THOMAS, Ref. 29, p.62.
32. J.H. WILSON, BSC Tinplate Research, S. Wales, unpublished work.
33. B. FOGG, *Sheet Met. Ind.* Feb 1967, 95.
34. N. WILTSHIRE, BHP's Role in Tinplate R and D; its Commercial Impact, First International Tinplate Conference ITRI, London, 1976.
35. LOUIS SCHULER, *Metal Forming Manual*, Gottingen, Wuertt. 1966.
36. J.A. GRAINGER, *Presswork and Presses,* Machinery Publishing Co. Ltd., Brighton, 1961.
37. D.F. EARY and E.A. REED, *Techniques of Pressworking Sheet Metal*, Staples Press, London, 1960, Hawthorn Books Inc., USA.
38. D.J. CAMPION, Tooling for Deep-drawing and Ironing, *Sheet Met. Ind.,* July 1976, 20.
39. DAVID CAMPION, Deep Drawing and Ironing — Theory and Practice, *Sheet Metal Ind.,* Feb 1980, 111; April 1980, 330; June 1980, 563; September 1980, 830.
40. P.B. MELLOR, Deep-drawing and Redrawing of Thin Sheet Materials, *Sheet Metal Ind.,* Dec 1977, 1180.
41. W. PANKNIN, C. SCHNEIDER and M. SODEIK, Plastic Deformation of Tinplate in Can Manufacturing, *Sheet Metal Ind.,* August 1976, 137.
42. T.Z. BLAZYNSKI, *Metal Forming: Tool Profiles and Flow,* MacMillan, London, 1976.

FURTHER READING

GEO E. DIETER, *Mechanical Metallurgy,* 2nd Ed. McGraw-Hill, New York, 1976.

T.Z. BLAZYNSKI, *Metal Forming: Tool Profiles and Flow,* McMillan, London, 1976.

G.W. ROWE, *An Introduction to the Principles of Metal Working,* Arnold, London, 1971.

W. JOHNSON and P.B. MELLOR, *Plasticity for Mechanical Engineers,* van Nostrand, London, 1962.

JOSEPH DATSKO, *Material Properties and Manufacturing Processes,* John Wiley, New York, 1966.

JOHN A. WALLER, *Press Tools and Presswork,* Portcullis Press, 1978.

H. MIAKELT, *Mechanical Presses,* (translated by R. Hardbottle), Arnold, London, 1968.

G. OEHLER, *Hydraulic Presses,* (translated by F. Minden), Arnold, London, 1968.

W.F. WALKER, *Guide to Press Tool Design,* Butterworth, London, 1970.

Chapter 6

Protective and Decorative Coating Systems

6.1 INTRODUCTION

The application to metal containers of organic coatings for protective and decorative purposes has been practised for a long time. The sale of branded products developed significantly during the last century, and this led to demands for specific external decoration: their use increased rapidly with the development of mass production methods and improved distribution to a growing area of demand.

In a similar manner, growth in the range of food products packed in metal containers and the need for longer storage periods in more difficult environmental conditions led to a demand for more effective internal protective systems.

Most of these were originally applied to the flat sheets of plate ("in the flat") before forming into containers (e.g. with soldered side seams).

Since then, great technological changes have taken place and are likely to continue. New and improved techniques such as two-piece can manufacture (DWI and DRD), welding of side seams and beading of can bodies for greater rigidity, together with wider use of alternative sheet metal to tinplate, have demanded more sophisticated materials; these demands have been reinforced by the introduction of more severe regulations covering health and safety and air pollution; and the increasing costs of raw materials and of energy.[1] The Food and Drugs Authority (FDA) in the United States, the European Economic Community Directives, and similar safety regulations in most countries impose strict rules, and these call for sophisticated formulation and application followed by careful product testing. A basic coating system will consist of:

One or two coats of lacquer on the surface intended for the can interior.

Externally: a coat of size and, when the canmaking and filling processes are demanding, a base coat or a base white ink, followed by several ink passes to form a print design, and finally a varnish coat. (Occasionally, an external colourless or golden-coloured lacquer is used to complete the protection and decoration of the external surface.) Normally, each coat will be stoved after application (except when specially formulated varnishes are applied "wet on wet" over inks).

Amongst recent developments, however, alternative and faster methods of curing or stoving are being used, for example by ultraviolet radiation, and by development of resin formulations which need lower temperatures and shorter curing cycles. These developments are discussed in section 6.4.

Although the development of the DWI can has led to a gradual change from lacquering and printing of sheet to that of fabricated cans, nevertheless a substantial amount of tinplate is still treated in sheet form; the increasing amounts of electrolytic chromium/chromium oxide coated steel and blackplate being used normally have to be lacquered and printed in sheet form.

6.2 PROTECTIVE COATINGS

6.2.1 Materials

Most coatings are supplied as solutions or dispersions of resinous materials in appropriate solvents, and require to be dried after application; they dry by solvent removal, oxidation and/or heat polymerisation. This is usually achieved by heating in a tunnel oven at up to 210°C for times up to 15 minutes, according to type and temperature employed.

The basic properties required of these materials are that they have adequate resistance to scratching and steam processing, and are free from extractable toxic constituents and any danger of imparting off-flavours; all aspects must be carefully tested after complete stoving. They must also on occasion be capable of withstanding severe fabrication conditions (beading, deep drawing, high temperature soldering). The type of internal coating, termed lacquer or enamel, will be related to the method of fabrication employed as well as to the conditions of use.

Early lacquers were invariably oleoresinous types, made from natural gums and resins, blended with drying oils such as dehydrated castor and tung. White spirit type solvents are common thinners for this class. They were and are suitable for a range of foods e.g. fruits and vegetables; even though many other types have since been developed, oleoresinous formulations are still extensively used, but on a reducing scale, because their relatively high solids content makes them economical and they are less critical for film weight control than many of the more modern synthetic products. The stoved lacquers will vary in thickness normally from 2 to 15 microns, according to type and end use; in practice, they are expressed in terms of weight per unit area (usually g/m^{-2}) as this gives a more practical method of measurement.

Many newer types of resins have been developed over recent years, having improved properties in particular respects (e.g. phenolic, epoxy, vinyl and acrylic). Not all of the newer drying media, however, are suitable for use internally on food cans. Amongst these are the alkyds, which are widely used in external decorative systems and the domestic decorative field; they are easy to modify for particular purposes, and are economical to use. However, they are not sufficiently free from taste to be used as internal can linings. However, the number of synthetic resins which provide bases for excellent internal coatings is still large, but as this covers a complex field, a detailed account of the numerous varieties is outside the scope of this book. Detailed descriptions are generally readily available from major manufacturers and in published literature.[2-6] Newbould has summarised the resins used nowadays as related to their end use; this is given in Table 6.2.1.[7] Reference 8 contains a similar analysis, as shown in Table 6.2.2, of coatings and products related to can types. It is of interest to note that some (e.g. vinyl copolymers and acrylics) can be pigmented to a white or pale colour after stoving, to provide an attractive interior for particular food packs. Epoxy-phenolic types can be formulated to contain a meat-release agent to ease removal of solid meat packs.

Generally, lacquers are formulated for particular types of products; for example, highly coloured fruits (red plums, raspberries, blackcurrants, etc.) are highly acidic and, as processing at 100°C after filling and closing the can is adequate, a lacquer having a high temperature resistance is not a major requirement; its main functions are to prevent the pigments present in the fruit from reacting with tin, thus avoiding unacceptable colour changes, and to prevent corrosion of the container leading to unacceptable metal pick-up by the product, or even perforation of the container. The lacquer must possess sufficient flexibility to avoid lacquer fracture during fabrication, which would result in acid attack on exposed metal. In the case of vegetables, some of their sulphur-containing proteins are liable to decompose during processing the filled can, which could result in the formulation of tin sulphide and black iron sulphide on exposed metal sites. Although these are not harmful, their appearance is unacceptable, and must be avoided, either by grinding fine zinc oxide or zinc carbonate into the lacquers or by using less permeable phenolic lacquers. The zinc compound will react with the evolved sulphur compounds to give innocuous white zinc sulphide. Because meats in general require a longer sterilising treatment than vegetables, the lacquers used for the latter in general are not adequate for meat products and need to possess more heat resistance while retaining film flexibility. Thus each class of product requires particular lacquer consideration, and the development of improved lacquers to withstand the more severe conditions of modern processes is continual. As mentioned earlier, beverages (beers and soft drinks), wines, etc., particularly require special lacquer types and different methods of application.

TABLE 6.2.1. CONVENTIONAL RESINS AND THEIR END USES

Coating	Resin types	Advantages	Disadvantages	End uses
Sizes	Alkyd Epoxy-amino Vinyl	Good flexibility Food substrate and intercoat adhesion	High volatile solvent content	As size coatings where improved fabrication required
Enamels	Styrenated-alkyd	Cheap Good-fair flexibility Good processing resistance	Poor chemical resistance Poor colour retention	Low flexibility non-process and processing cans
	Polyester	Good flexibility Good colour retention Good process resistance	More expensive than alkyds	Processing and non-processed cans, caps, closures Deep drawing caps Aerosols
	Acrylic	Good flexibility Good colour retention Good process resistance	Odours Expensive	Processing and non-processed caps Processing bodies
	Epoxy-ester	Good product resistance Good hardness	Limited flexibility Expensive	Toothpaste tubes
	Vinyl	Very good flexibility Good processing resistance – if modified	Low solids Very expensive Thermoplastic unless modified UV/heat degradable	Deep drawing caps Toothpaste tubes Drawn processing cans

continued on page 198

Coating	Resin types	Advantages	Disadvantages	End uses
Varnishes	Alkyd	Cheap Good hardness	Fair processing resistance Poor colour retention Flexibility fair	Low flexibility non-processing cans
	Polyester	*As Enamels*	*As Enamels*	*As Enamels*
	Acrylic	*As Enamels*	*As Enamels*	*As Enamels*
	Epoxy-ester	Good hardness Good-fair flexibility	Expensive Fair colour retention	Screw caps Crown corks
	Vinyl	*As Enamels*	*As Enamels*	*As Enamels*
	Epoxy-amino	Good-fair flexibility Good colour retention Good processing resistance	Expensive	Caps and ends
Lacquers	Phenolic	Good produce resistance Excellent sulphur resistance Good processing resistance	Very poor flexibility Poor plate wetting	Can bodies
	Epoxy-phenolic (EP)	Good product, sulphur and processing resistance Good flexibility	Expensive Lowish solids content	Can bodies and ends Internal and external non-compound caps
	Epoxy-ester phenolic	Good flexibility Cheaper than EP	Only fair product sulphur and processing resistance	Can bodies and ends
	Epoxy-urea formaldehyde	Good product and processing resistance Alcohol resistance good Cheaper than EP	Only fair drawing properties beverage cans	Internal spray lacquers for beer
	Vinyl	Excellent flexibility Good alcohol resistance	Low solids content Expensive UV/heat degradable Monomer thought to be carcinogenic	Internal spray lacquers for beer beverage cans
	Organosol	Excellent flexibility and adhesion product resistance Excellent compound adhesion	Only fair flow-out Low process resistance High film weights required Monomer thought to be carcinogenic	High flexibility cap and closure linings Deep drawing lacquers
	Polybutadiene	Very cheap Good product resistance	Odour Only fair flexibility	Beer can bodies
	Oleoresinous	Cheap Good product resistance Good sulphur resistance, with ZnO or $ZnCO_3$	High baking schedules required	General food can bodies

TABLE 6.2.2 STEEL CAN COATINGS

Coating & products	Can type	Comments
Acrylic Applied as liner for cherries, pears, pie fillings, single-serve puddings, some soups and vegetables Acceptable to aseptic packs	For all exteriors, and interiors where a clean, white look is desired. As a high-solids side seam stripe for roller and spray application to wire welded cans Clear water coat for D & I cans Waterborne, spray liner for 3-piece beverage cans	Expensive, but takes heat processing well and is suited to water-borne and high-solids coatings. Employed primarily on can exteriors because of flavour problems with some products. Makes an excellent white coat and assures colour retention when pigmented
Alkyd (Exterior use)	Quart oil can ends in clear gold or aluminium pigment Aerosol domes in white enamel Flat sheet roll-coat as decorating as a high-solids white base coat and varnish	Low cost, used monthly as an exterior varnish over inks, because it would present flavour and colour problems inside the can. Trend is towards supplanting conventional alkyds with polyesters which are oil-free alkyds
Epoxy amine Beer and soft drinks Dairy products Fish, ham and sauerkraut Nonfoods such as furniture polish, hair spray and paint	D&I as beer and beverage base coat Draw/Redraw Can ends Over-varnish on aerosol cans and domes	Costly, but has excellent adhesion, colour, flexibility and imparts no off-flavour, scorch resistance and abrasion resistance. Used in interior or as varnish and and size coat. Employed now in water-borne coatings and, with polyamide, as a side seam stripe in high solids form for welded cans
Epoxy phenolic Beer and soft drinks (as a base coat) The coating for food, including fish, fruit, infant formula, juice, meat, olive pie fillings, ravioli, soups, spaghetti, and meat balls, tomato products, vegetables	All steels and cans	Big in volume for can. interiors. Used in Europe as a universal coating. Main attributes are steam processing resistance, adhesion flexibility and imparts no off-flavour. Especially suited for acid-aggressive products. Has excellent properties as a base coat under acrylic and vinyl
Epoxy-Phenolic with Zinc oxide/carbonate metallic aluminium powder For sulphur containing foods: fish, meats, soups and vegetables	3-piece tin plate can ends 2-piece single draw cans (307 × 113)	Used primarily to prevent tin sulphide straining, flexibility and clean appearance are the main attributes of these "C" enamels.

continued on page 200

Coating & products	Can type	Comments
Oleoresinous Beer and soft drinks (as a base coat) Fruit drinks A wide variety of fruits and vegetables*, including acid fruits	All except draw/redraw	A general purpose, gold-coloured coating least expensive of all. When additional protection is required, it can serve as the undercoat for another lacquer. Can be used in both high-solids and waterborne systems *when lacquer contains zinc oxide
Phenolic Acid fruits, fish, meats, pet food, soups and vegetables Nonfoods	All steel cans where coating can be flat-applied	Low-cost, exceptional acid resistance and good sulphur resistance. Film thickness restricted by its inflexibility Tendency to impart off-flavour and odour to some foods.
Polybutadiene Citrus plus soups and vegetables (if zinc oxide is added to the coating) Beer and soft drinks	Restricted to 3-piece can bodies because of its poor fabricality	Good adhesion, chemical resistance and ability to undergo processing are its chief characteristics. Resistance properties and cost comparable to the oleoresinous class
Sulphide stain-resistant oleoresinous – "C" enamel (has zinc oxide) Asparagus, beans, beef stew, beets, chicken broth, chocolate syrup, corn, peas, potatoes, spinach, soups, tomato products and sulphur-containing foods	Soldered and welded tin plate Can ends	Low cost, flexible, often used as a top coat over epoxy-phenolic. Not for use with acid fruits.
Vinyl Beer and soft drinks Fruit drinks Tomato juice Nonfoods	Draw/Redraw, even in triple draw cans, for coating is very formable Roller-applied topcoat for beer and beverage	Flexible and off-flavour-free (used for years, with beer). Resistant to acid and alkaline products. Notsuitable for high-temperature, long-processed meats and foods. Big as a clear exterior coating.
Vinyl organosol Foods	Draw/Redraw Topcoat on beer and beverage can ends 3-piece food can ends Thin tin plate which has to be beaded	Especially suited for thicker film application (9-10 mg. per square inch) Good fabricability, superior corrosion resistance. May prove ideal fo welded cans, which allow more severe beading. Still used in low-solids, solvent-borne manner.

6.2.2 Application and curing

Three basic methods are used for the application of these materials:

1. roller coating;
2. spraying, or
3. dipping (almost obsolete).

By far the bulk of all coating application is carried out on roller coating machines; this includes the coating of material in sheet and coil form (in the latter case reverse roller coating is used, and is described later), and of cylindrical can bodies externally. Spraying techniques are used mainly to coat the inside surface of can bodies, the production of which has shown a substantial increase since the development of DWI and DRD techniques. Application by dipping into liquid coating is confined to a few special cases.

Roller coating

The principle of the roller coating machine is shown in Fig. 6.2.1 and the photograph of a modern coater in Fig. 6.2.2. It consists of a series of rollers which successfully pick up a quantity of the material to be applied from a tray and apply it uniformly at a controlled thickness over the surface of the metal sheet fed through the machine.

The thickness of wet lacquer applied to the sheet is governed mainly by the width of the gap between the fountain and transfer roller, therefore this distance can be set precisely. It is also vital, in order to achieve adequate control of the film weight, that the fountain and transfer rolls, especially the latter, run truly concentrically and have been ground accurately round and smooth. The pressure between the transfer and application rollers and on the sheet during passage can be finely adjusted to ensure uniform film thickness and avoid visual irregularities.

A small gap will exist between the back edge of one sheet and the front edge of the succeeding sheet, which will allow the coating material to be transferred to the pressure roller; to prevent this being deposited on the under surface of the following sheet, a thin scraper is fitted which effectively removes it to the return tray; the blade must be maintained in good condition for effective scraping. Some coater designs incorporate a mechanism which drops the pressure roller away from the application roller when a sheet is not passing through the assembly.

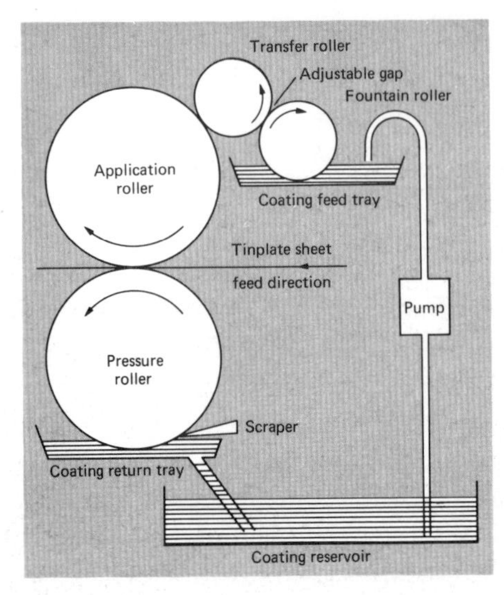

FIGURE 6.2.1 ROLLER COATER DESIGN

Roller coaters are used to apply lacquers, pigmented coatings and varnishes. The machines must be capable of applying "stencilled" coatings — certain areas of the sheet not being coated, i.e. left plain — to permit, for example, soldering or welding of the side seams of cylindrical can bodies. The application roller consists of a steel roll covered with a suitable rubber-faced blanket, or a moulded rubber or glue–glycerine composition; it can take many different forms. Stencil channels are accurately cut out from the roll covering by any suitable method, e.g. a lathe-type machine designed for this purpose.

The roller coater is incorporated into a large production line, shown schematically in Fig. 6.2.3, consisting of

(i) an automatic feeder and run-in conveyor;
(ii) a roller coating machine;
(iii) a continuous over.

The sheet feeder

Sheet metal is delivered in a bulk package in which normally 1000-2000 sheets are stacked on to a wooden platform. This is run into the feeder which, by means of suction cups, airblast and magnetic separators, picks up and passes each sheet forward separately to the coating machine. The sheet is registered both in the "front-to-back" and side-to-side directions by various mechanical means, so that it enters the coating machine squarely to allow precise location of any clear areas or "stencil". Precise location is also required when feeding sheets to a printing machine for accurately registering print images.

FIGURE 6.2.2 ROLLER COATING MACHINE

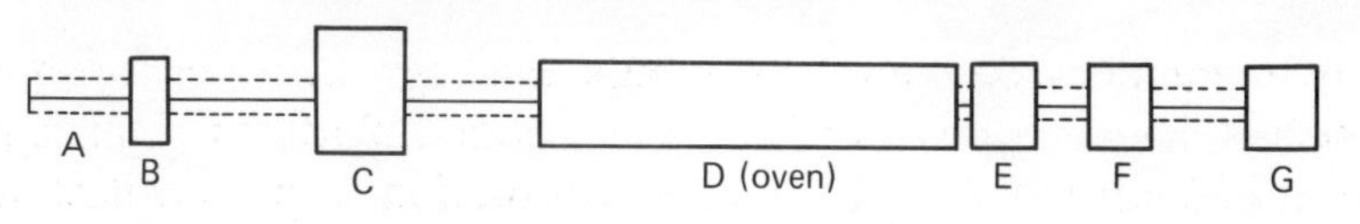

A Run-in conveyor
B Automatic feeder
C Roller coater
D Oven
E Sheet cool
F Unloader
G Piler

FIGURE 6.2.3 SCHEMATIC LAYOUT OF TYPICAL LACQUERING LINE

The roller coater

Machines can operate at speeds up to 7000 sheets per hour, but this will depend on plate quality (especially size and gauge uniformity near to the edge) as well as machine design. Thin sheet edges can be dented at higher speeds, which could lead to registration problems; sheet size and "geometry" must also be maintained within the accepted tolerances to avoid stencil faults.

FIGURE 6.2.4 OVEN ENTRY WITH CONVEYOR WICKETS

The stoving oven

On emerging from the roller coater (or printing machine) the sheets are fed into the stoving oven through which passes a continuous conveyor: this is fitted which metal frames or "wickets" at 18 mm to 25 mm pitch. As the coated sheet enters the oven, it is picked up by a wicket and transported in a near vertical position through the oven (illustrated in Fig. 6.2.4.). Oven lengths range up to 35 m, and a full oven can contain up to 1500 sheets, giving a total load on the conveyor up to 2 tonnes. Oven length will relate to line speed and pitch of the wickets. The oven will usually have external heating chambers, fired with gas or oil, and air is blown through the chamber and into the oven, the hot air being directed between the sheets by a ducting system. Part of it is recirculated, and the remainder exhausted into the atmosphere. An alternative heating arrangement, used occasionally, is a row of burners along the base of the oven and the hot atmosphere circulated by means of fans. Close control of oven temperature is vital and a target of ±2°C is aimed for, so a good thermostatic control system is required. Stoving schedule is defined as the minimum time over which peak sheet temperature is held: Fig. 6.2.5. illustrates a typical temperature

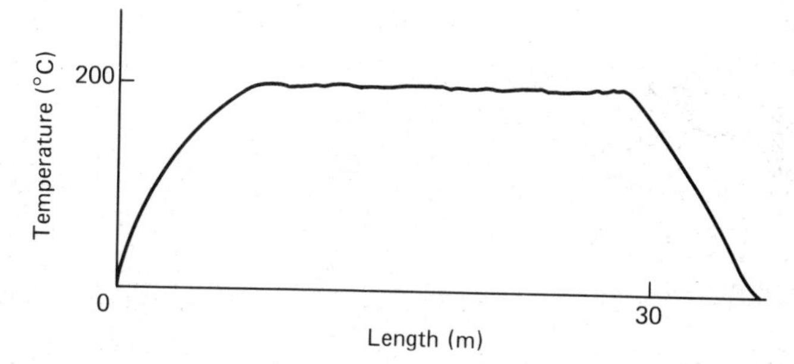

FIGURE 6.2.5 TYPICAL OVEN TEMPERATURE CHART

chart. The front part of the curve represents a come-up zone in which the temperature is usually raised as rapidly as possible. The next, and longest, zone should provide a temperature of up to 215°C (for a varnish/print oven the temperature desired is up to 180°C, and is seldom below 120°C. This oven will be shorter than that required for lacquers, perhaps only 10 m to 15 m long). At the end of the stoving period, shortly before the sheets emerge from the oven, they are cooled rapidly by means of powerful air fans, so as to eliminate the thermoplastic and after-tack stage to avoid any danger of set-off and sheet sticking when the sheets are automatically restacked at the end of the line. The term "no after-tack" is used to indicate that the coating must not possess any trace of tackiness at room temperature, following stoving, to avoid any possibility of the coating surface being picked off by, and transferred to, the next sheet when the plate is stacked into a pile and thus under a heavy weight. Also, even slight sheet sticking is to be avoided, as it will affect the smooth feeding of the sheets during subsequent high-speed coating and printing operations. If the reverse surface of the plate is then to be coated, the sheets will be reversed automatically on the run-out to the restacking unit; alternatively, the stack of plate can be turned over as a whole in a bundle turning unit.

6.2.3 Testing of lacquers and enamels

Complete testing of internal can lacquers is a complicated process and will be described only broadly here. Testing can be divided into three stages

(i) evaluation of the suitability of the lacquer for storage and application on a coater of other apparatus;
(ii) its behaviour as a cured film on the metal surface in the relevant fabrication processes;
(iii) resistance of the cured lacquer to the specified service conditions.

Physical characteristics of lacquers

Lacquer is usually characterised by its solids content, specific gravity, viscosity as determined by an agreed method, and its flash point, together with no unacceptable changes in these characteristics during storage. It is standard practice to specify acceptable ranges for these characteristics, and generally a batch of an established lacquer will be capable of being applied satisfactorily if its physical properties are within the range laid down.

In the case of a lacquer under development, application trials will need to be carried out on a production machine, as it has not yet been possible to develop laboratory tests which will adequately reflect full-scale application behaviour. A further precaution often adopted is to test application and stoving on several batches of plate, including a variety of grades, tincoating masses and surface finishes.

The stoved lacquer film will be examined visually for wetting behaviour, any evidence of pinholing, "flecking", and uneven application. It is essential to check that the film weight applied and the stoving conditions (time and temperature) are well within the standards. Film weight may be checked by accurately punching discs of know diameter from positions adjacent to each side edge of the sheets and determining the loss in weight on removal of the lacquer in a hot caustic soda solution, in suitable mixed solvents or electrolytically in, say a 2% sodium carbonate electrolyte using a high current density. Film weight is recorded in terms of grammes per square metre.

Attempts have been made to develop laboratory tests to assess degree of cure, but none of these is wholly satisfactory. This characteristic is included in the testing described in the next section.

Behaviour of the film during fabrication

Although there are several laboratory tests in use to assess lacquer film behaviour on drawing, beading, etc., and they are useful as a means of weeding-out the unacceptable from the promising, they are not entirely reliable; recourse has to be made to production-type equipment for final assessment. These units will need to be expertly set up and run at normal speeds.

Fabrication testing can be divided into three types:

(a) lacquered built-up cans and components for food products and beverages;
(b) lacquered drawn, DWI and DRD cans and components for similar products;
(c) lacquered drawn or built-up cans and components for non-food products, i.e. those classified as "general line".

After appropriate forming, the components are tested in several ways.

Loss of adhesion and lacquer failure: this is more likely to occur at more heavily deformed areas and is examined visually and also be immersion in acidified copper sulphate solution; dirty brown metallic copper will be deposited on faulty sites. Attempts have been made to develop more precise methods of evaluating these features, but none, as yet, has shown any real advantage over the practical types described above. Many other methods are used to evaluate such aspects as: abrasion resistance, surface lubricity, heat/scorch resistance, etc., and reliance has still to be placed on these empirical tests in the absence of a more fundamental approach (electrolytic tests are, however, commonly used to check for minute metal exposure in spray lacquered beverage cans, e.g. the widely used Enamel Rater test). A more fundamental approach to this type of evaluation requires separate testing of many factors, such as adhesion, cohesion, flexibility and extensibility, which contribute to lacquer behaviour during fabrication. Efforts are continuing to develop improved laboratory techniques. None has been found to be entirely satisfactory.

Resistance of the lacquer film to service conditions

Food cans are subjected to severe conditions of high temperature for up to several hours when the food contents are sterilised after filling; they will subsequently be stored for comparatively long times under a range of temperatures. Lacquers applied to beer and soft drink cans will also be required to withstand similar severe conditions.

The usual methods of testing will involve manufacturing containers from the test lacquered plate, filling them and processing appropriately, and finally assessing the container quality after various periods of storage. Food cans will be stored at 37°C and at ambient temperatures for up to 2 years. Filled beverage cans will be stored at somewhat lower temperatures and for shorter periods. Synthetic products may be used to give more rapid results, and are useful in screening new lacquers.

It is essential to examine for any deleterious effect of the product on the lacquered plate and of the cured lacquer on the product. Examination will include the following:

On the lacquer

(i) softening, blistering or stripping,
(ii) sulphur staining of the metal surface (normally by meats, fish and some vegetables),
(iii) "dezincing" of sulphur-resistant (SR) lacquers by vegetables,
(iv) "greening" by some vegetables, particularly spinach and cabbage,
(v) corrosion of the metal at fractures of the lacquer film.

On the product

changes in flavour, colour, clarity of the fruit syrups or liquor; tin and iron contents.

Organoleptic testing of beverages and some food products for off-flavour and odour is a specialised technique requiring trained personnel and the use of isolated rooms, and of a "hidden" as well as a declared control. Analysis of trace elements to less than 1 ppm is frequently carried out; it demands the use of modern analytical techniques and equipment.

Testing of "general line" cans is usually confined to periodic examination of filled cans after various storage periods. A complication results from the extremely wide range of product types that can be packed; often a representative "short list" is selected to reduce the tremendous effort that would otherwise be needed. In this case, strip testing consisting of immersing narrow strips of the lacquered plate in the products contained in stoppered small glass containers is very useful; this simplified method is not considered reliable with food products.

Test procedures are usually agreed with customers, and lacquer suppliers also will co-operate closely with material testing; joint development is encouraged and indeed vital.

In addition to the importance of sheet gauge, shape and dimensions, its surface characteristics can also have a marked bearing on surface wetting and lacquer adhesion. The light lubricant film applied at the end of the metallic coating process must be held strictly within the agreed limits, to ensure acceptable coverage; excessive amounts of lubricant will give rise to severe dewetting, to a degree also dependent on the characteristics of the lacquer being applied. The nature of the passivation treatment given to the surface, subsequent to electroplating, can markedly affect adhesion of the cured lacquer; each new lacquer should be tested carefully against the various types of passivation available, to ensure an optimum performance.

6.3 DECORATIVE SYSTEMS

6.3.1 Materials

As mentioned in section 6.1, the purposes of decorating a container are varied, and will depend particularly on its type and application. Primarily, it is aimed at substantially improving the container appearance, to assist its marketing appeal; when required, the design will include a description of a branded product. It will also significantly improve external corrosion resistance.

Some examples of particular requirements are:

OT food cans (with soldered or welded side seams) — adequate process resistance

Drawn containers — to withstand the deformation involved without damage or adhesion loss

Aerosols — stringent product resistance

Caps and closures — withstand drawing followed by threading without surface damage or adhesion loss

Trays — frequently a difficult forming operation; requiring good resistance to abrasion and alcoholic drinks.

The external decorative system is similar in many respects to the protective type used internally, in that the constituents are generally dispersed in volatile solvents; applied on roller coating machines (apart from the printed image) and stoved in long "tunnel" ovens.

Where severe deformation of the metal is involved in the fabrication process, the system is likely to consist of a size, a base coat, a printed image containing several colours and a final varnish film.

The size will be mainly epoxy or vinyl types, but some are based on alkyds. They possess good adhesion to the plate surface, although the precise mechanism of this is still not fully understood. They are white, pale shades or clear lacquers so as to avoid interference with the print colours; they make an important contribution to the fundamental properties of the decoration and must be capable of being stoved without residual tack. i.e. completely dry with no trace of tackiness, and must not discolour on further stoving. Only thin films are applied, about 1.5 g/m^2 (up to 1 mg/in^2), but this does involve a coating operation and a pass through the oven. Continual attempts are being made to improve adhesion characteristics without this costly operation, and base coatings having improved adhesion directly to the plate surface have been developed; the use of size-free systems has increased in recent years in cases involving limited forming.

Roller coatings in the main will be white or a pale colour, to provide a high-quality printed design. As they must possess good opacity, film weights of the order of 15 g/m^2 are generally used; excellent flow characteristics are also required, and the coatings also must dry completely on stoving without residual tack, so that stacking can be done at speeds of 100 per minute. Again, no colour change on any subsequent stoving is acceptable. Gradual improvement in opacity, together with brighter cleaner whites, have reduced film weights needed; adoption of polyester whites has been of real help in these improvements. Acrylic resins in particular have resulted in coatings being developed with much improved heat stability. Both of these are especially useful for the filled food cans requiring steam processing at high temperatures.

The application of base materials is an important but expensive part of metal decoration. Specifically it requires high energy levels to achieve a full cure, and material costs are significant. It is now, however, possible to cure white coatings at temperatures of 115°C compared with 160-170°C a few years ago. In all roller coating applications it is important to control the application viscosity, usually by solvent addition or temperature adjustment.

As the printing operation is more complex than the application of roller coatings, even wider developments have taken place, especially towards newer techniques; these are described later. Inks for the conventional off-set printing process have been improved steadily in colour strength, in closer control over rheological properties to provide better flow and transfer, and in print quality; improving stability during steam processing has also been achieved. A particular achievement has been the development of new formulations based on lead-free pigments covering a complete range of shades which conform to the British Toys (Safety) Regulations; these specify maxima for lead and arsenic contents, and also of soluble antimony, barium, cadmium and chromium.

Finishing varnishes are required to protect inks and base coatings from damage during fabrication and later use, and to add gloss to the print image (some designs specify a matte finish for which a special varnish would be used). They need good flow properties as they are applied on roller coaters, their "wet on wet" property is important; and again must stove hard without residual tack. A good resistance to high-temperature steam processing is required when used with processed food cans, and to alcohol when applied to waiter trays.

Attempts have been made with some limited success to eliminate final varnishing; its elimination calls for considerable improvement to ink quality in terms of resistance to marring and scuffing during feeding operations, and to various products.

Technical Publications 33—Organic Coatings for Metal Containers,[9] and TC 32 Metal Decorating,[10] both by Metal Box plc, although published in 1964 give a useful basic account of decorative materials used at that time. More recent developments are presented by Lott[11] and Holt.[12] Additional information is available in publications issued by major coatings and ink suppliers in the United Kingdom, e.g. Holdens Surface Coatings Ltd., IPIC Paints, Packaging and Coil Coating Division, Coates etc.; in the United States, e.g. Dupont, Glidden-Durkee, Inmont, etc.; and in most other countries.

The newer materials developed for use with new printing processes will be described later under those headings.

6.3.2 Traditional offset lithography

This technique has been used for over a century for decorating sheet metal; it is carried out by an offset process, the ink being transferred from a printing plate to an offset blanket, and then on to the substrate.

Conventional offset lithography employs a planographic metal printing plate having hydrophobic areas in the form of the desired design receptive to printing ink, and the surrounding hydrophilic

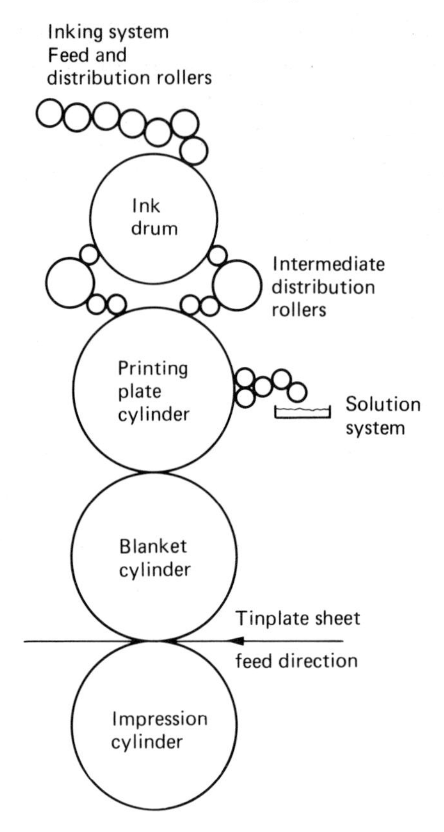

FIGURE 6.3.1 PRINTING PRESS DESIGN

area receptive to water. A uniform coating of ink is applied to the printing plate but is retained by the design area only; uniformity of ink is ensured by a complex system of (16 or more) steel rollers between the ink duct and the printing plate. Figure 6.3.1. illustrates in section a typical tin printing press; two of the three main cylinders carry respectively the printing plate and a rubber blanket; the third is an impression cylinder which causes the metal sheet to bear on to the blanket; this ensures that the sheets pick up a uniform image of the print. A separate smaller set of damping rollers applies a thin uniform film of fountain solution to the printing plate to ensure that ink is not picked up by the non-image print areas. Maintaining a careful balance between the quantities of ink and solution calls for considerable skill on the part of the printer; this is essential to ensure good definition and the colour target with minimal variation. The quality of the image laid down on to the metal sheet is dependent on many factors, such as shop temperature and humidity, and the effects of the applied solution on the characteristics of the ink. Due to the offset process, the maximum amount of ink transferred is limited to approximately 2.5μ, i.e the inks must have high colour strength. Single colour machines were originally the norm, each colour needing to be stoved before the next could be applied. Sometimes a two-colour machine, termed a Y printer, due to its design, as shown in Fig. 6.3.2 was used if the various colours needed for the design did not overlap. The Y machine is an expensive unit and has not been widely adopted.

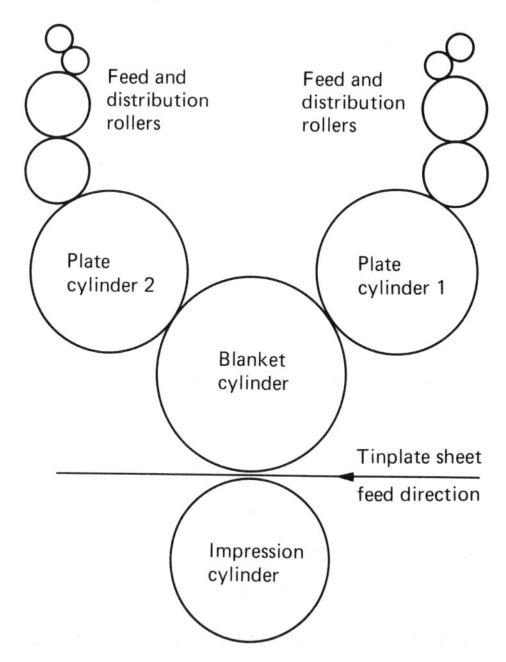

FIGURE 6.3.2 Y-TYPE PRINTING PRESS

As a number of colours can be required for many designs, attempts have also been made to print one wet ink on to the previous wet ink ("wet on wet") by adjusting the inks' tack properties, so that the one already applied is not "picked off" by the blanket during the second operation. However, this method is not wholly satisfactory for high-speed, high-class printing, as it cannot be completely controlled by the printer. An alternative method of reducing the number of ink passes needed for the design is to apply a second ink over a stoved first colour, thereby giving a third shade.

Many methods of "setting" the ink in between single colour presses in tandem have been investigated: these have included high-temperature air blasts, flame treatment and other physical methods. The most widely used is the application of UV radiation to the print; the UV unit is small, so that it is readily fitted between printing decks in a conventional printing line, and the system has a very short curing time, about 0.01 second. Setting inks by this method particularly calls for a new approach to ink formulation, and considerable success has been achieved.

Developed of the UV curing method as a separate technique is described later.

The most effective means of reducing the number of passes required has come about through a better understanding of the various factors which control trapping of inks "wet on wet" on a non-absorbent substrate; ink film thickness and tack are particularly important. This has resulted in extensive use of two-colour printing presses in tandem, without the need for intermediate stoving. As a result, a design requiring two colours can be produced in one line pass.

Stoving of the decorative systems is carried out in similar tunnel-type ovens to those used for protective lacquers, but they are normally shorter, usually about 16m (55ft). They must be capable mechanically of transporting the sheets smoothly through the oven at speeds of up to 6000 sheets per hour. Peak stoving temperatures will also be lower, normally ranging between 120° and 180°C.

6.3.3 Reproduction

The printing process applies a single colour at a time, and hence the original design has to be separated out into the minimum number of colours that will reproduce satisfactorily the original. This separation and reproduction is a complicated operation, much of it done either photographically or electronically.

The design to be reproduced by the printing process will be a photograph, painting, or sketch produced by studio artists for the purpose.

Simple designs may consist of and be reproduced using a small number of specially selected inks, but any pictorial design actually consists of hundreds of shades. To reproduce these, special techniques are used. The basic colours — frequently yellow, magenta (blue/shade red), cyan (green/shade blue) and black — are separated and then reproduced using a halftone technique. This technique produces an image using a screen which breaks the design up into dots, typically 150 dots per inch would be used. The darker the area to be printed, the larger the dots and the lighter (e.g. in tint areas) the smaller. Control of dot size and quality is a critical aspect of high-quality printing and reproduction art is a skilled part of decoration.

Printing plates are selected for their quality of printing, life and cost. They vary from presensitized aluminium to complex metal combinations, but all have the facility to form a grease-receptive ink area and a water-receptive non-image area. They may be prepared from positive or negative films, according to the plate/process; platemaking can be fully automatic operation. Printing plates are typically about 1000 mm long by 900 mm wide, and hence many images can be laid out on one plate. The actual layout is produced using a "step and repeat" machine. When decorating a very small component, e.g. a screw cap, there can be several hundred images on a single plate. To facilitate subsequent stamping or forming, the layout must be precise. The life of a plate is important in two respects: it must be adequate for a specific run, but frequently it may be reused for subsequent production runs. Multi-metal plates, e.g. steel, copper, chrome, are used for the latter due to their robustness and reliability. General accounts of platemaking are given in references (10), (12) and (13).

6.4 NEW MATERIALS AND PROCESSES UNDER DEVELOPMENT

Extensive development of materials, protective and decorative, has been proceeding for a long time, and this will continue. Over recent years improvements have been directed at radically new types of material, together with new methods of application and curing. These changes present a complex picture, but a very useful summary has been presented by Newbould[7] which classifies them into four distinct groups:

Group 1: Same application technique/same curing method as conventional coatings.
Group 2: Same application technique/different curing method.
Group 3: Different application technique/same curing method.
Group 4: Different application technique/different curing method.

Group 1

This is subdivided into three types: water-borne coatings, high solids coatings, and non-aqueous dispersions (NADs). Efforts on the development of these types have been intense, because the industry has been faced for a number of years with increasing demands for reduction in air pollution due to organic solvents, and in the United States, the Environmental Protection Agency (EPA) declared in 1976 that all solvents must be regarded as potentially photoreactive. This abolished the earlier separation into "exempt" and "non-exempt" solvents, and was replaced by the concept of limiting total organic emission of volatile organic compounds (VOC). The total emission permitted varies with the end use of the coatings; it ranges from 0.31 kg/l to 0.66 kg/l of coating (excluding any water content). Detail is given by Guerrier,[1] and many other countries have introduced, or are debating, similar legislation.

Group 2

Several new methods of curing have been examined, the two most promising of which are ultraviolet (UV) and electron beam (EB) curing. The former has been used for a number of years in several packaging activities, firstly in several paper and plastics applications, and more recently in the exterior decorative fields; in view of its increasing importance, it is described in some detail later.

Electron beam curing is not yet so highly developed in packaging, although it is by no means new; it was first evaluated in the 1950s. Basically the electrons are generated from a heated source and accelerated *in vacuo* towards the target. The beam can be focused and is used either as a narrow beam scanned across the coated metal component, or as a wide "curtain".

Direct flame curing and infrared heating are other possibilities, and both in fact are already in use in particular applications, e.g. external coatings on round metal containers. Induction curing has promise as it provides very rapid heating; up to 375°C in a few seconds is possible.

Group 3

The possibilities of applying water-borne coatings by electrodeposition have been examined periodically over many years. Earlier water-reducible resins were usually anionic, due to the presence of carboxyl groups used to provide water solubility. When a can is made the anode, the ionic coating will migrate to it and be deposited on uncoated areas; subsequent stoving will cure this coating in the usual way. It is a particularly useful technique for the repair of internally coated cans, thereby giving cans of a very low metal exposure standard. More recently there has been a move towards cathodic materials.

For some purposes the use of powder coatings provides a useful technique; electrostatic spraying is usual. Internal coating of large drums requiring heavy coatings can be successfully carried out, but for the thinner coatings used for metal containers the powder needs to have a small particle size to ensure film continuity. Special precautions have to be taken to prevent fine powder from becoming electrostatically charged, as this would lead to flow problems and blocking of the spray equipment. Many resin types have been found to be suitable, both thermosetting (acrylic polyester and epoxy) and thermoplastic, such as nylon, polyvinyl chloride and polyethylene, give a good performance. The surface of beverage can welded side seams are coated commercially in one instance using powder coatings.

Group 4

Many hot melt coatings, based on microcrystalline waxes, have been evaluated, particularly as alternatives to vinyls, as internal beverage can coatings. They are heated and then applied by spraying on to cans at room temperature, but to avoid rough coatings being formed the cans have subsequently to be heated to a slightly elevated temperature. These materials tend to give off-flavour problems as they are prone to absorb odours from the environment, and are inferior to vinyls in this respect. Their application may also be restricted because much higher film weights than is usual for vinyl types are required to provide adequate resistance to product–can reaction. They are also affected by the high temperatures often present in pasteuriser units.

Coil coating methods

"Coil" (strip) coating of aluminium has been employed for a number of years, generally applied by aluminium manufacturers rather than the fabricators. The coated coil is used for drawn cans, and caps and closures, as well as for decorative purposes. It has only recently been used for tinplate strip, at speeds up to 230 m per minute. Curing temperatures up to 400°C are sometimes employed for aluminium, giving very short curing times, less than 60 seconds. As tin melts at 232°C, curing is at up to 230°C for times of at least 5 minutes and the full economy of coil coating is not achieved. Some more recent developments suggest that momentary melting of the tin coating on tinplate does not cause undue harm, and trials have been carried out at temperatures of 450°C over a few seconds only. These temperature limitations would not apply to chromium/chromium oxide coated nor to uncoated blackplate.

The coating materials are usually applied in roller coaters similar to those used for sheet, but some lines use the reverse roller coating technique as, it is claimed, it deposits a smoother, more uniform, film. In principle, the application roller rotates in the opposite direction to the strip movement. As contact with the wet coating must not occur between application and complete stoving, the strip must be maintained under high tension. Water cooling after stoving is sometimes employed to accelerate the point at which recoiling can be done.

Application of continuous coating to tinplate in coil form has not been extensive; in order to compete economically with sheet lacquering, high strip speeds and large outputs are needed.

Internal can spraying

As the amount of metal deformation taking place in the DWI operation is considerable, with substantial disruption of its surface, no pre-applied lacquer film can be subjected to it without damage; the can body must therefore be coated internally after fabrication. This has called for development of a very precise technique to provide the standard of protection required for packing beverages, especially soft drinks, as exposed iron is anodic to these corrosive acid beverages. A high standard of coating is assisted by the absence of a side seam, always a source of potential failure.

As the DWI lines operate at speeds of about 800 cans per minute, rapid, precise, automatic spraying machines are vital; normally each line will contain about three stationary or oscillating spray gun assemblies. Reid[(5)] has described the development of a system of this type, using static spray guns to meet the line speed requirements. These are fitted with nozzles designed to give fine sprays under high lacquer pressure (i.e. without air pressure, which tends to form blisters) and uniform placement; the nozzles "beam" at various rates to allow for distance to and contour of the can surface, so that uniform coatings can be achieved. These stationary nozzles require more precise setting than the reciprocating versions. A parallel lacquer development was also required — in terms of viscosity, solvent balance, temperature, pressure and spray times — to achieve the high standard of lacquer coverage required. The lacquer supply is held within a closed circuit, which includes a pump, heater and precise timer.

As with three-piece cans, electrochemical tests are used, together with chemical metal exposure tests, to monitor can quality in production.

New printing techniques

Dry Offset

A development from lithography which is used extensively in three-piece food can production is termed dry offset (alternatively know as offset letterpress). Printing is by means of a relief wrap around plate, i.e. the image is in relief and hence water is eliminated. As these plates do not require dampening, several advantages result:

> The need for a precise balance between ink and water does not arise, so start-up is much quicker, with less waste, fewer stoppages and greater throughput, and a significantly albeit small reduction in consumption of expensive ink.

The plate, however, is many times more expensive than the litho plate, and is not so suitable for single short runs. Print quality also is not to the same standard; its use for multicolour halftone designs is severely limited. Correct machine setting is more difficult than with offset litho, and ink rheology also plays a more important part. In spite of these shortcomings, on balance the process is favoured in many less critical applications.

Container printing

Printing of formed cylindrical containers ("in the round") has in fact been used for some years, e.g. rigid and collapsible aluminium extruded tubes on individual printers applying up to six colours. In this work, the containers are roller coated with white on rather simple machines and stoved on a "peg" conveyor through a hot-air oven or infrared heated oven, followed by printing and similar stoving; for higher class work they would be varnished in a final operation. All colours are applied from each printing plate to the same printing blanket, and the whole multicolour image is transferred to the tube in one operation, i.e. in one rotation of the tube.

Although attempts are usually made in designing the decoration to keep the individual colours slightly separate, this is difficult and will restrict the scope of design. The conditions for satisfactory trapping of wet inks over each other on the blanket are opposite to those required for satisfactory transfer in the printing operation, but, by careful control of rheological properties and of thickness applied, some progress has been made towards printing multicolour halftone designs. With automatic tube transfer and improvements in machine design speeds in excess of 150 per minute have been achieved.

With the development of the D and WI beverage cans, high-speed can printing machines have been designed, initially in the United States, capable of handling over 800 cans per minute. They can apply up to four colours, followed by a varnish coat. The can surface may be sized or have had a white base applied on a high-speed can coater and stoved before printing. It is claimed that, as the cans have been rigorously cleaned to remove the lubricant used in wall-ironing, adhesion of organic coatings on to their surface is generally better than on tinplate. The machine shown in Fig. 6.4.1

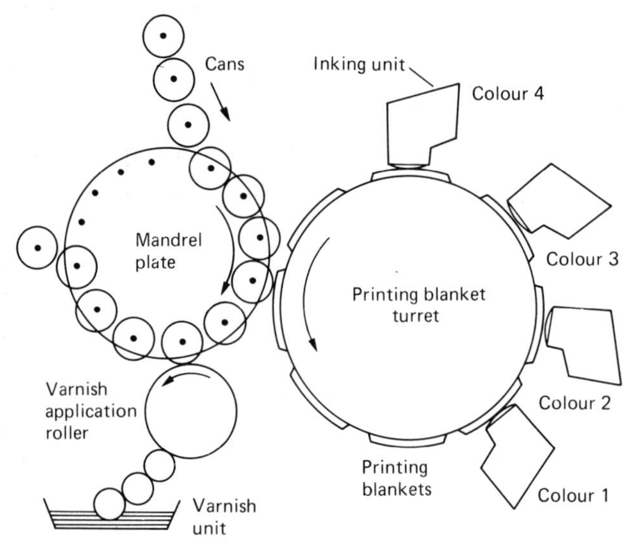

FIGURE 6.4.1 TWO-PIECE CAN DECORATOR

has four printing inks and eight printing blankets. In addition to rotating, the cans are also moving along a circumferential path during printing; uniform printing pressure is ensured by means of a cam controlled motion. Transferring the cans at high speed, on to and off the moving mandrels on which they are supported during printing, has been achieved by a complex mechanism. The four printing inks apply the image to the blanket by means of a dry offset plate, and all four colours making up the design are applied to each of the eight blankets. A varnish may also be applied, finally, before the whole decoration system is stoved. Thus the designs that can be applied suffer the same shortcomings as those mentioned under tube printing earlier. Ink properties must therefore by controlled very accurately; it is essential that their consistency and rheological properties remain constant over the high roller speeds obtaining in this process, coupled with inevitable temperature variation. The varnish properties also must be adequate at high speed. Attempts are continually being made to improve ink properties so that some "over-printing" whilst still wet can be achieved.

UV radiation curing

The development of UV radiation curing technology probably represents the most significant advance in coating practice for many years. Although it has been in use as a curing technique for some years with paper and plastic materials, substantial effort has been needed to enable it to be used extensively in the metal decorating field.

Formulations amenable to photocuring will consist of several ingredients:

(a) The major portion — up to 70% by weight — will contain a mixture of polymers; these will have a material effect on the physical properties of the coating.

(b) A reactive monomer which functions as a cross-linking diluent. It is usually liquid and non-volatile; the need for light colour, low viscosity and high purity favours acrylic compounds.

(c) The photo-initiators and sensitizers will be the active ingredients of the formulation; usually selected from the aromatic ketones, xanthone, benzoin and their derivatives.

Curing is very rapid, under appropriate UV radiation; when the photosensitive molecules absorb this radiated energy they release free radicals which polymerise or cross-link the liquid into a solid coating. Further, as the substrate is not heated, it can be handled immediately for further operations. Thus, a substantial energy saving results — about 50% — over heat curing; space requirement is substantially reduced giving faster production throughput, and the need for expensive tunnel ovens and equipment is eliminated. It can also be the means of reducing the number of passes needed in multicolour printing, as UV curing of a base coat and of inks between presses avoids the problems associated with "wet on wet" printing and can offer four, five or six colour printing lines. Printing "wet-on-dry" will always ensure better quality. Several commercial units are available for fitting between the decks of multicolour units or on the ends of tin printing lines.

The principal disadvantage is the very high cost of UV materials; inks are typically double the cost of conventional inks.

External coatings having thicker films can suffer from incomplete penetration of the radiation to the substrate surface, due to being scattered or absorbed by the added pigments.

Ultraviolet curing is currently, therefore, limited to printing rather than base coating.

These materials might require approaching twice the normal radiation dosage; further development in this area is also required, coupled with a reduction in the cost of irradiation generation.

UV sources

Many types of lamp are now available. Originally the high-pressure mercury arc lamp was used. It operated at 80 W/cm (200 W/in) and ranged up to 1 m in length; the tube was of quartz, as this allowed maximum transmission of UV radiation and was suitable for operating temperatures up to 800°C. Lamp life was guaranteed to be not less than 1000 hours, and the output did not drop by more than 15% throughout the life.

A detailed cost analysis of UV in comparison with thermal curing drawn up by the Celanese Chemical Company was reproduced by Lindstrom[(14)] in 1976. It was based on use of a coil coating system.

After a slow start, use of UV curing materials has become firmly established, and it will grow steadily. Pressure on methods of reducing atmospheric pollution is likely to increase, but economic factors will also be important, especially the relative cost of materials. Some technical problems will need to be solved; more complete reaction is essential if the need for post-stoving also is to be avoided. More efficient longer-lasting lamps are still needed, coupled with their integration into printing units designed for this particular purpose.

REFERENCES

1. J. GUERRIER, The Changing Scene of Tinplate Coatings and Lacquers, Holden Europe SA, Paper no. 37 of the Second International Tinplate Conference, ITRI, London, 1980.
2. T.G. GREEN *et al.*, Organic Protective and Decorative Coatings for Metal Containers, *Bull. Inst. Metal Finishing.* **4,** (3), (1954), 227.
3. Organic Coatings for Metal Containers, TGG, Metal Box plc, Technical Communication No. 33, 1964.
4. ARTHUR ST. CLAIR, The Coating of Beer and Beverage Cans, *Packaging Review,* **93,** (6), (1973), 51.
5. R.T. READ, Recent Developments in Protective Finishes for Metal Containers, Part 1, Internal Organic Coatings, *J. Oil Col. Chem. Assoc.,* **58,** (1975), 51.

6. S.E. CANNING, The Significance of Developments in Canmaking Techniques: Protective Lacquers, Pira Seminar, PIRA, Leatherhead, 1976.
7. A.J. NEWBOULD, The Role of New Technologies in Coatings for the Packaging Industry, *J. Oil Col. Chem. Assoc.*, **60,** (1977), 53.
8. Steel Can Coatings Undergoing Major Changes. Report by the Committee of Tin Mill Products Producers, *Am. Iron and Steel Institute Metal Finishing,* Oct 1981, 39.
9. Organic Coatings for Metal Containers, TC 33, Metal Box plc, 1964.
10. Metal Decorating, TC 32, Metal Box plc, 1964.
11. A.D. LOTT, Recent Developments in the Printing and Decorating of Metal Containers, *Tin International,* July 1973, 229.
12. J.C. HOLT, Recent Developments in Protective Finishes for Metal Containers. Part II: External Finishes, *J. Oil Col. Chem. Assoc.* **58,** (1975), 57.
13. S. KAPEL, Advances in Modern Metal Decorating, *Tin International,* June 1981, 208.
14. R.S. LINDSTROM *et al.*, Radiation Curing: Cool, Quick and Clean, *Machine Design,* 6, May 1976, 96.

Chapter 7

Corrosion of Cans

7.1 INTRODUCTION

Low carbon steel is particularly susceptible to damage by corrosion in a very wide range of atmospheric and aqueous environments and the effects of small amounts of sulphur dioxide, H_2S and organic and inorganic acids can be such as to increase the rate of destruction by several orders of magnitude. The presence of electropositive metals such as copper, and the establishment of local reactive sites as a result of cold work by bending and forming, will also give rise to accelerated corrosion. In all cases, the speed and extent of corrosion depend on the establishment of areas at different electropotentials — anodic and cathodic areas between which there is a flow of electric current. These are most conventionaly described as galvanic cells set up between the anodic and cathodic areas; the former are electronegative to the cathodic areas and the most damaging situation will obtain when a large current is generated between a large cathode and a small anode. Anodic polarisation will accelerate the conversion of a metal into its ions, which are removed if they are soluble, and so a thin section can be perforated by solution or weakened until it is fractured under stress or pressure. The concentration of the soluble ions in the electrolyte will increase continuously as corrosion persists.

Tinplate, on which the surfaces consist of an almost continuous layer of pure tin, is highly resistance to atmospheric corrosion and capable of withstanding moisture and many materials in solution for substantial periods. Pinholes in the tincoating are specially damaging, as the tin is the cathode and the exposed iron the reactive anode; under severe conditions, very heavy localised corrosion can occur at pinholes and other breaks. The obsolescent hot dipped tinplate will, in addition to carrying heavier tin coating weights, have a substantial tin–iron alloy layer, $FeSn_2$, between the tin and steel base; the alloy layer on electrolytic tinplate is much thinner, being formed mainly during rapid flash melting of the electrodeposited tin coating. This is electropositive to the base and also to the tin, and acts as a chemically inert barrier to attack on the steel base.

In the case of food containers in particular, the permitted extent of corrosion is minimal, in terms of pick-up by the contents (although they only amount to a few parts per million), as this may cause changes in flavour and appearance of the product (e.g. loss of colour in fruit packs and some vegetables; haze in beer; etc.), discoloration of internal can surfaces, etc.

Galvanic corrosion mechanisms are generally studied by means of polarisation diagrams, in which increasing currents are applied to a selected electrochemical cell containing an appropriate electrolyte under aerobic or anaerobic conditions, and the resulting potential differences are measured. Their use for this purpose was first suggested by U.R. Evans some years ago, and considerably developed for metallic coating systems by Hoar;[1] the polarisation diagrams are widely known as "Evans diagrams" (Fig. 7.1.1).

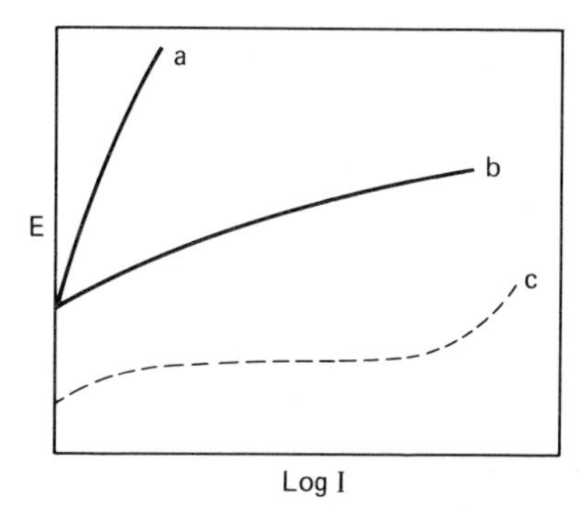

(a) anode polarisation curves: a. highly polarised; b. little polarised; c. anodic complexing

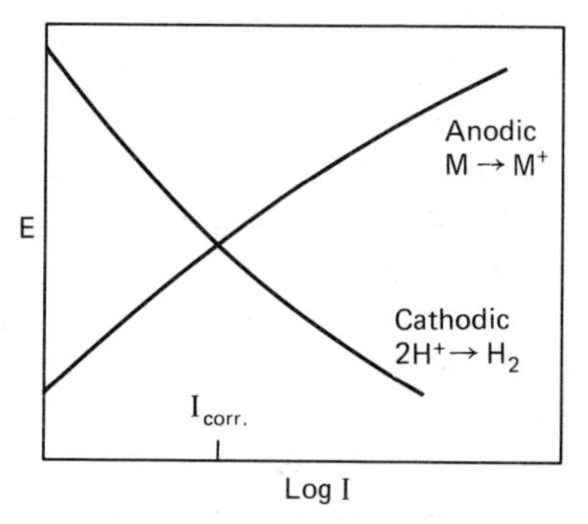

(b) anode and cathode polarisation curves; corrosion current given by the point of intersection

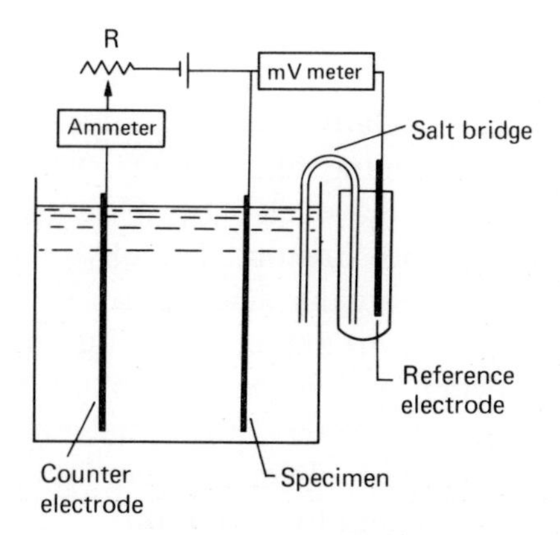

(c) a simple circuit for drawing a polarisation curve

FIGURE 7.1.1

Internal corrosion processes can be complicated by several factors:

1. The normal polarity of the tin–iron galvanic couple can, under certain circumstances, be reversed, the tin then being the anode and thus sacrificial to the steel base; in that case the tin will be dissolved, so exposing larger areas of steel which are subsequently rapidly corroded.
2. The corrosion rate is greatly affected by the extent of porosity, giving access through the tincoating to steel base, as well as by the structure of the intermediate $FeSn_2$ alloy layer, and to a lesser extent the passivated "oxide" layer and the oil film on the plate surface (these were described in Chapter 2, section 2.12).
3. The corrodibility of the low carbon steel base will vary appreciably according to its composition (especially copper, phosphorus and sulphur contents) and metallographic structure, and to the pickling and annealing practices.
4. The corrosivity of the packed product can vary appreciably according to the type of product, and also from batch to batch of the same product; a particular example of the latter is red plums.
5. The environment will also be affected by complexing agents (e.g. organic fruit acids) and the presence of accelerators or inhibitors. Residual oxygen in the packed can will also affect the corrosion rate.

Gabe discusses[2] the conditions under which polarity reversal occurs; stannous ions can be complexed by food products containing tartrates, citrates, etc., and he indicates a method of calculating the conditions for reversal. Relative polarities also depend on hydrogen overpotential, tin developing high values, whilst iron does not. Studies of these effects have also been described by Kamm,[3] Willey,[4] Sherlock[5] and Britton.[6] Scully[7] also discusses the complexing of tin ions by organic acids. Patrick[8] gives a very useful review of the internal environment obtaining in unlacquered cans packed with acid fruits, and shows that many more complexants than the usual fruit acids (oxalic, citric, malic and tartaric) can be present, and have similar effects. Other substances shown to complex tin are chloride and hydroxyl ions.[1,9,10]

Undoubtedly, many as yet untraced accelerators and inhibitors can be present in some products (discussed in more detail in section 7.2). Gabe[2] summarises the possible corroding conditions under four categories; internal and normally anaerobic, without complexants, anaerobic with complexants, external and aerobic with complexants and the more usual aerobic conditions without complexants. The standard electrode potentials are calculated for solutions of unit activity, and if this is lowered by the presence of a complexant, the potential becomes negative. A substantial reduction in activity can so change the value that the polarity of a cell with another metal may be reversed. In the tin–iron cell, where the standard potentials are −0.136 and −0.44 volts respectively, this can result in selective corrosion of one or the other, according to the composition of the atmosphere and the solution. Depolarizers decrease the extent of polarization at any given current and so increase the maximum corrosion cell current. Taking the four categories of exposure:

1. *Anaerobic conditions without complexants.* Tin is normally cathodic, but may be anodic if depolarizers are present. The rate of tin solution is slow, but the amount may become objectionable. Beer cans are internally lacquered to ensure a very low level of metal exposure.
2. *Anaerobic conditions in the presence of complexants.* Tin is anodic and will protect the iron until completely removed. This is the normal condition for citrus fruit and similar packs.
3. *Aerobic conditions without complexants.* The tin is cathodic and exposed iron rusts rapidly. Normal atmospheric corrosion.
4. *Aerobic conditions with complexants.* Corrosion rates are high in the presence of oxidant (depolarizer) feedstuffs; these will include green vegetables and alkaline detergents in aerosol packs.

Metallography has developed into a useful method for the study of coating corrosion; it can monitor progress by surface and taper section study, from inside and outside.

Transmission and scanning electron microscopy are also of considerable use in conjunction with electrochemical measurement, and analysis of the products. These have been of particular help in clarifying the role of the $FeSn_2$ alloy layer.[11] Earlier work by Britton *et al*[12] and Covert *et al*[13] had shown that it was cathodic to both tin and iron in deaerated organic acid solutions, and that attack on the alloy, whether coupled or not, was very slight. Thus, under the anaerobic conditions obtaining within processed food cans, tin is anodic to the steel, and will be dissolved sacrificially at pore sites; but the barrier effect of the inert alloy layer will prevent a significant increase in the steel cathode area.

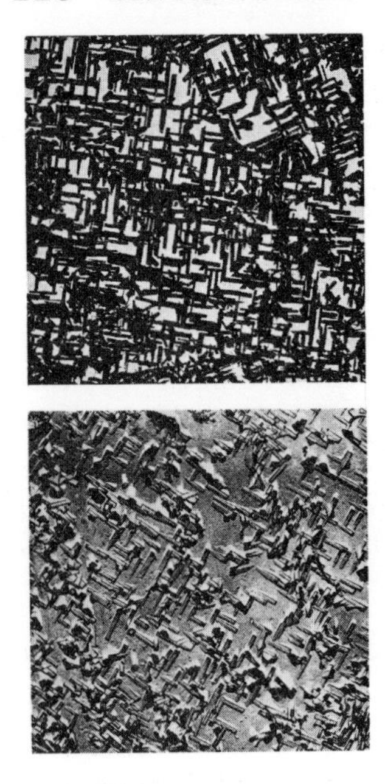

FIGURE 7.1.2 ALLOY LAYER OF ELECTROLYTIC TINPLATE (x 16000)

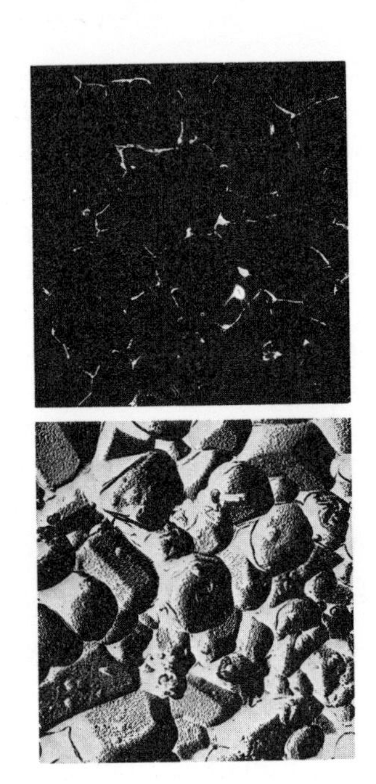

FIGURE 7.1.3 ALLOY LAYER OF NON-DIPPED TINPLATE (x 3000)

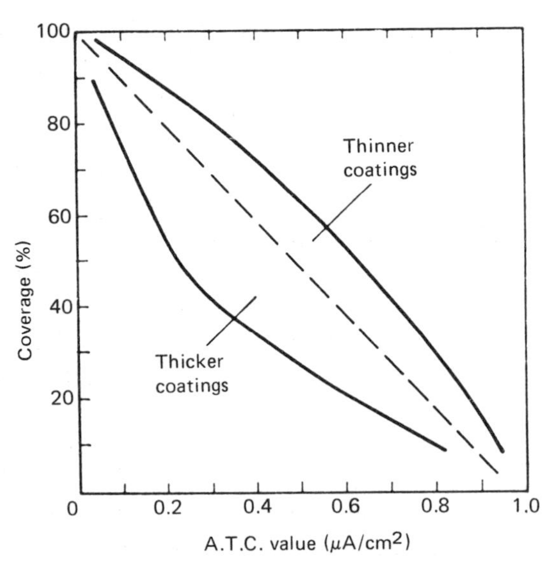

FIGURE 7.1.4 RELATIONSHIP BETWEEN ATC VALUE AND ALLOY CONTINUITY FOR TINPLATE

Thus, as had been suspected for some time, the density — or degree of continuity — of the alloy layer has a material effect on rate of corrosion. Its continuity has been measured by point counting from replica micrographs by Gabe and Beard;[14]

Discombe *et al*[15] showed (Figs. 7.1.2 and 3) by isolating the alloy layer and viewing it directly in transmission that similar results were obtained. The alloy–tin couple (ATC) test, described in the next section, had been developed by Kamm[16] earlier as a rapid method of measuring the corrodibility of electrolytic tinplate in grapefruit juice. Gabe demonstrated a correlation between ATC values and continuity (Fig. 7.1.4), as measured by point counting from replica micrographs.

7.2 CORROSION PROCESSES AND FACTORS AFFECTING THEM

7.2.1 Acidic fruit products

Normal practice is to pack in a way which preserves the natural colours, normally using internal lacquering for red fruits, such as plums and cherries and many berries, and unlacquered tinplate for the so-called "white" fruits — citrus, peaches, pears and apples. The lacquer will be special to the fruit, and the actual procedure may depend on local custom and practice. Efficient coverage of metal, where specified, will be ensured by applying two internal lacquer coats "in the flat", and a stripe along the side seam during bodymaking.

The general pattern of corrosion in the lacquered can is very different from that when there is none, because the only exposure of metal is at scratches in the lacquer coats and at cracks along the heavily deformed side seam. There is virtually no protection from anodic behaviour by the tin, and the principal mechanism is solution of the steel base by fruit acids. The corrodibility of this base then assumes great importance. In the plain, or unlacquered, can the controlling reaction will be the solution of tin. This metal gives initial protection to the steel base, but is removed at a rate controlled by the cathodic reaction on exposed steel. When the removal has exposed a large area, the rate of attack on the base will depend on the extent to which the alloy layer is present as a protection.

In both cases, however, metal will be taken into solution and hydrogen gas will be evolved and the ultimate "shelf life" will depend on whether the hydrogen pressure builds up to such a value as to cause the ends of the can to bulge, this is harmless but indistinguishable from one "blown" due to bacterial spoilage, or the metal content of the contents reaches an unacceptable level. The amount of hydrogen generated is less than the stoichiometric equivalent of the metal dissolved, because some diffuses through the steel of the can.

The total environment inside the container is extremely complex, with effects arising from particular growing conditions for each fruit, including the use of fertiliser and insect-killing sprays and chemical aids to growth and ripening. For example, the Victoria plum exhibits a wide variation in corrosivity, depending on the locality in which it is grown, seasonal conditions and the details of husbandry practised by the growers. Even the sulphur content of some sugars and copper from cooking vessels may require to be controlled.

Tinplate characteristics

The coating weight is a major factor influencing the rate of corrosion, but this must also be related to the method of manufacture, because this influences the extent of alloy layer formation. The alloy layer is more complete on hot-dipped tinplate than on electrolytic, and in the early years of the use of electrolytically coated tinplate its performance was, at best, erratic. As electrotinning is the only method of bulk production used nowadays in the industrialised countries, much knowledge has been gained on the control of alloy layer formation and on its influence on corrosion behaviour. A comprehensive account of this work is given in *Technology of Tinplate*[(17)] and in references (18) and (19).

Considerable work has also been carried out on the influence of composition and structure on corrodibility of the steel base. Copper, sulphur and phosphorus were found to be particularly deleterious, and American canmakers set the following limits for their tinplate specifications:

	Cu	P	S
For most corrosive products: Type L	0.06%	0.015%	0.05%
For moderately corrosive products: Type MR	0.20%	0.02%	0.05%

Other countries have adopted similar steel grades, and in the United Kingdom the Cu maximum for acid fruits has been set at 0.08%.

Early work appeared to indicate that Si in excess of 0.05% was damaging, but recent research indicates that it can be up to 0.10% in the above specification.

This has also led to the development of rapid tests for corrodibility and significant improvements in steel base resistance to attack. The four tests most widely used are the iron solution value (ISV), the pickle lag value (PLV), the tincoating grain size (TGS) and the alloy–tin couple current (ATC); these are referred to sometimes as "Special Property" tests.

Initially, electrolytic tinplate for "plain" fruit cans was supplied to a specification based on steel chemistry, TGS, ISV and PLV.[20-22] This specification was found inadequate to ensure a consistently high shelf life of filled cans. Later development of the ATC test and its incorporation in specifications, together with the application of a heavier electrolytic tin coating, raised the performance level to that of hot-dipped tinplate. The $FeSn_2$ crystallites in hot-dipped tinplate are nodular in shape and based on a very large number of nucleation sites, while those generated in the electrodeposited coating by subsequent heating are acicular, exhibiting marked epitaxial growth on the iron crystals of the base. These acicular crystals give a less protective coating than that provided by the nodules. Electrolytic tinplate produced from a sodium stannate bath was found initially to have better and more uniform corrosion behaviour, and it was described as grade A ET plate when conforming to the following criteria:

ATC value <0.12 μA cm^{-2} (average >0.05 μA cm^{-2})
PLV value <10 sec
ISV value <20 μg Fe
TGS ASTM nonferrous grade 8 (or larger)

As a result of improved control and practice, electrotinplate was then produced from stannous sulphate (Ferrostan) and stannous chloride baths to meet the same criteria. This is described as type K electrolytic tinplate.

ATC values for electrotinplate from acid baths can, in exceptional cases, range up to 0.25 μA cm^{-2}, but attention to the following points ensures that a high proportion will be within the type K limits in all four tests:

1. The steel is of "fruit can" grade type L or MR containing less than 0.020% P and not more than 0.08% Cu, with very few large non-metallic inclusions, particularly carbide particles.
2. The pickling conditions are controlled within close operating limits, and the annealing is preferably carried out in an HNX atmosphere containing 4-8% H_2 and very low moisture and oxygen contents.

3. Preplating cleaning conditions on the tinning lines are also maintained within narrow laid down limits.

Experience has confirmed that plate meeting this specification gives improved shelf life with citrus products, especially grapefruit juice. The performance with other types of fruit packed in plain cans has been varied, some reports indicating substantially improved shelf life, while others find no improvement over good "run-of-the-mill" tinplate. In the author's experience a large tin grain size and a low ATC value are the essentials for a good shelf life.

If the cans are lacquered, there is still a significant correlation between the four parameters and the shelf life, even when the steel composition is within the specified limits. Experience in Britain indicates that the combination of suitable steel composition, and acceptable pickle lag value, with freedom from massive cementite (nowadays a rare occurrence) gives best results. All of these must be combined with minimum metal exposure in the finished can.

Differences in specification will be found between locations, sources of plate, and also type and nature of product packed. The number of fruit cans produced and therefore the quantity of plate with "Special Properties" required, varies considerably between countries, with a substantial increase from Northern Europe to Southern Africa.

7.2.2 Surface staining

Many food products, notably meats, some vegetables, fish packs, and milk, can cause staining of the tincoating. This can range from general to localised and vary in colour from deep blue to almost black. Although the precise nature of the coloured films is not known, they are generally classified as sulphide stains. These are not harmful, but their appearance is often objectionable. Iron sulphide deposits can also be formed at sites of damage, particularly along the side seam; these are not adherent and can fall off on to the product.

The temperatures used to sterilise the can contents may cause breakdown and evolution of sulphur-bearing volatiles from protein compounds, coupled, in the case of meat products, with the meat acids, reacting with the tin surface. It can be largely avoided by the use of sulphur resisting (SR) lacquer into which quantities of zinc oxide or carbonate are added before being applied to the plate surface; these react with the sulphur-bearing gases to form almost invisible white zinc sulphide. Sometimes lacquers wholly imperm-eable to the vapours are used. Anodic surface treatments like the

Protectatin, in which the surface oxide layer is considerably increased, can be very effective in reducing staining; but as the layer is subject to severe local damage during canforming, the protective processes must be carried out after canmaking. As they require treatment times up to 15 minutes, followed by efficient washing and drying, their cost is too high for most applications.

Effective avoidance of tin sulphide staining can result in the "unused" sulphur vapours causing heavier iron sulphide deposits being formed at damaged sites.

The nature of meat, its age and curing treatment can also have a significant effect on susceptibility to causing staining.

As minimal lacquer damage and adhesion loss are required, the nature of the passivation film is particularly important. Generally, those carried out cathodically only (not containing an anodic stage) give a better performance. Close control of oil film weight is essential. Optimum lacquer types also need to be carefully selected.

Some canned milk products can also give rise to different types of surface staining; these are especially dependent on the type of passivation used and also on the surface damage caused to the film during canmaking operations and changes brought about at soldering temperatures. Again, the nature of the milk and canning conditions can affect the extent of staining. Preferred passivation treatments in this case also are the cathodic-only type in sodium dichromate solution; the precise composition of the passivation film (amounts of tin oxides, chromium metal and chromium oxides) can be particularly important. One, based on sodium carbonate, is favoured by some milk packers. Use of lacquered plate is sometimes specified, but does not always eliminate all forms of staining.

A particular form of staining has arisen in recent years with lacquered tinplate ends in cans of several milk-based puddings; investigations of its causes has been described by Harden and Walpole.[(23)] This form of corrosion was shown to be dependent on many factors (pack formulation, composition of headspace gases within the can, type of lacquer and lacquering conditions) in addition to the characteristics of the tinplate. Electrochemical studies coupled with x-ray photo-electron spectroscopy showed that underlacquer detinning in creamed rice is greater when high chromium levels are present in the passivation film and when significant $FeSn_2$ alloy is exposed at the metal surface. An electrochemical "alloy exposure" test was adopted to evaluate the tinplate surfaces.

7.2.3 Carbonated beverages

As with many acid fruits and other foods, the corrosivity of carbonated beverages packed in tinplate cans varies widely from the relatively inert types to some extremely corrosive varieties. Carbonated soft drinks are essentially formulated mixtures of flavoured syrups and carbonated water, containing flavours, acidulants, colours and sometimes other ingredients. Their variation in performance were experienced many years ago when the three-piece can — consisting of a body having two lacquer coats and the soldered seam side-striped with lacquer, and closed with two lacquered ends — became established for beverages. Daly,[24] in his general paper dealing with most aspects of internal and external can corrosion, reported that the behaviour of soft drinks depended to a considerable extent on the presence of Azo dyes (Amaranth), nitrates, copper, and the amount of oxygen remaining in the filled can. In addition to careful consideration of product formulation, it was recommended that all formulations should be test packed before commercial marketing; this involved small-scale filling and storage of cans up to at least 6 months, with examination of trace metal pick-up, organoleptic features and the appearance of the can's internal surface. As this extensive testing is both costly and time-consuming, many attempts have been made to develop short-term laboratory tests which will provide realistic indications of probable shelf life and corrosion behaviour; Daly describes a Corrosivity Tester developed at that time (1959).

Extensive work has continued since that time to develop a more complete understanding of the corrosion mechanisms in canned soft drinks, and the effects of plate factors, product, packing practice and storage temperature. Koehler *et al*[25] studied these aspects in detail, and concluded that the extent of corrosion could be significantly reduced most effectively by:

adopting means of reducing the corrosivity of the product (careful control of the use of Azo-type dyes, coupled with good cannery practice (low residual air contents, low copper, nitrite control, etc.);

and decreasing the metal exposure inside the can to a very low level.

The copper and phosphorus contents of the steel are important in that both, if present in appreciable amounts, can increase considerably the rate of corrosion. But type MR steel, which can have a copper content ranging up to 0.20% and a maximum phosphorus content of 0.02% was found to be satisfactory. Rephosphorised steel containing higher contents was used for a while to provide adequate end strength, but was replaced some time ago by plate made from nitrogenised steel, and nowadays a DR plate is generally used. Thus there is no basic problem in providing a steel appropriate to beverage can specification — copper and phosphorus less than 0.20% and less than 0.02% respectively. Further development of a Corrosivity Tester, particularly for the type of product where steel is anodic to tin, is also described in the paper.

Beese *et al*[26] have described their studies of the corrosion mechanisms in a range of carbonated beverages, including the test methods developed for these purposes. They classified canned soft drinks as follows:

Group 1 Beverages containing dyes or oxidants which tend to attack the tin directly; as they vary considerably the group is further subdivided into those in which tin is anodic to steel, and the type where steel is anodic to tin.

Group 2 Beverages which do not cause appreciable corrosion of tin, either by direct attack or by coupling to steel; the tin serves as an inert barrier, resulting in relatively low iron pick-up also. The Colas fall into this group.

Group 3 This includes the products in which tin is anodic to and protects the steel electrochemically; thus gradual detinning takes place at exposed areas, and rates or iron pick-up range from low to moderate.

Alderson[27] has discussed the importance of canning technology in relation to improved shelf life and container performance, and emphasises the role played by storage temperature and the restricted shelf life in tropical regions.

Although beers in general are less corrosive than the majority of soft drinks, similar product testing prior to commercial packing is essential, and equally low metal exposure levels of the fully lacquered cans used are maintained.

As a result of these findings and general experience, it is the usual practice for can manufacturers and packers to agree on a code of practice, which would cover the following points:

1. The formulation must be established prior to commercial packing.
2. Specified carbonation levels must be adhered to, related to the can size (in general, the smaller the can diameter, the higher the carbonation level which can be tolerated), coupled with specified headspace levels and residual oxygen contents.

3. Sulphite levels must be kept to agreed low levels.
4. Adequate treatment of water is vital, including filtration to remove colloidal material.
5. Trace metal (e.g. copper) contents in water.

Fruit juices are regarded in general as acid fruit products, and are packed normally into "plain" cans, even though hot-filling is normally adequate to ensure a microbiologically stable product. Unlacquered cans are normally used. The plate parameters discussed in section 7.2.1 are also of importance with fruit juices, particularly ATC values and tin grain size.

7.3 ATMOSPHERIC CORROSION

The principal external factors influencing the corrosion of metals are the humidity and composition of the atmosphere. In the case of tinplate containers, the weight of tin coating and the character of the external packaging will affect the corrosion resistance to a considerable degree.

The naturally formed surface oxide on tinplate is thin and has little effect in increasing corrosion resistance, but this can be improved by the "passivating" treatment and oil film application during manufacture. It is damaged very easily during forming operations and impact or abrasion during manufacture.

The influence of atmospheric humidity increases very considerably as the relative value rises above about 70%. If the tinplate surface is either contaminated by solid deposits, particularly if the particles are porous, or it is in contact with a porous fabric, such as paper or wood, the effect is exaggerated, particularly when acids are leached from these materials. The harmful effect is increased, possibly by orders of magnitude, when the atmosphere contains sulphur dioxide or the oxides of nitrogen. These dissolve to form acids which have considerable solvent activity. Chlorides are present in coastal and marine atmospheres and their presence, combined as it is with humidity cycling on cargo vessels, can cause rapid corrosion. Even the adhesives used to affix labels have to be checked for possible contributions to damaging conditions.

The tin coating is itself very thin (usually less than 1.5μm) and contains pores which expose the alloy layer and the steel surface. The number of pores will increase sharply if the weight of tin coating is reduced, and exposure of underlayers will be appreciably greater in areas of substantial deformation such as the expansion beads on the can ends.

The mechanism of external corrosion is complex. Tin is cathodic to iron under normal conditions and the galvanic cell set up between them at localised pore sites is of such a character that attack on the iron is accelerated by the presence of tin. Occasionally, the rust will spread, particularly on uncoated blackplate, along very narrow paths which are almost straight for short distances, but change direction randomly without crossing. This is known as filiform corrosion and is illustrated in Fig. 7.3.1. Appreciable rusting can also occur during the heat treatment necessary for sterilisation of the can contents. This is usually carried out in steam heated retorts at over 100°C, followed by water cooling to prevent over-cooking the contents. It is essential to ensure a low oxygen content in the retort by adequate venting, and to cool only to such a degree that the residual sensible heat is enough to dry off the exteriors of the cans. It is essential that this is complete before the filled sterilised cans are packed into cartons or other bulk packs. Corrosion may also result from spillage of acids or brine in the can contents, or the use of contaminated or chlorinated water for cooling. Alkaline water may etch the tin surface and water containing sodium chloride will cause accelerated attack. The position may be improved by diluting with high-purity water and/or chemical treatment as well as by using water as cold as possible in order to shorten the time of contact during cooling.

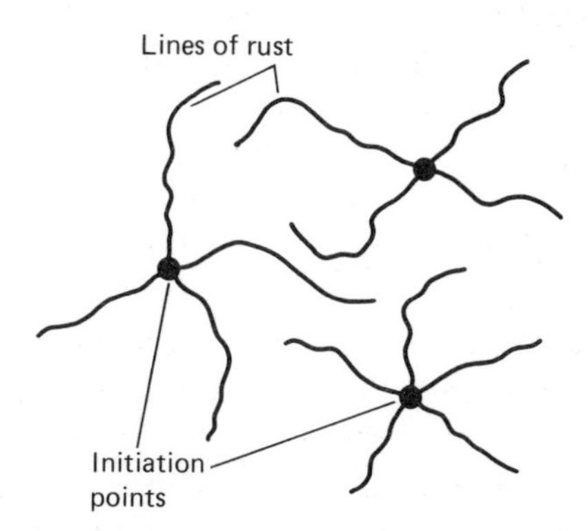

FIGURE 7.3.1 FILIFORM CORROSION

Rusting can also be caused by unsuitable conditions of transport and storage. "Sweat damage" due to humidity cycling, especially when the journey is from temperate to tropical zones. A consignment of canned goods loaded at a low ambient temperature can resin cool for a very long time, so increasing the liability to condensation and damage. This can be considered as a high probability on a voyage, for example from Europe to India, unless loading is carried out at an elevated temperature, and special precautions are taken to ensure efficient ventilation and low humidity. Several surveys of this kind of damage have been carried out. Duly[31] showed that a cargo tended to remain at its loading temperature throughout long sea voyages, and a study of the losses in canned milk transported from Vancouver to the United Kingdom was carried out by Seyer and Donald.[32] Cohen and Nelson[33] reviewed the liability to "sweat damage" in marine transport from the West Coast of the United States and found that it was most probable when loading took place between December and March. They recommended that

1. A film of corrosion inhibitor should be applied before labelling.
2. The bulk package of cans should be protected by inhibiting vapour barriers (with suitable outer packaging).
3. There should be improved methods of controlling the moisture content of the atmosphere in the cargo holds of ships.

Many of these recommendations apply also to the transport of tinplate and cans by rail. Special bulk packages and wrapping paper impregnated with a vapour phase inhibitor provide local environments, sometimes inside specially insulated cars, which are loaded under favourable ambient conditions. Warehouses are normally well insulated, and may be heated so as to eliminate condensation. This is especially important when batches of tinplate to be printed or coated are stored, since moisture condensation can cause dewetting of the coatings. Uncoated tinplate is liable to suffer extensive rusting if stored under adverse conditions. Special pamphlets have been prepared by tinplate and can manufacturers and their transport agencies, detailing the precautions to be taken during handling, transport and storage. Barbiere[34] describes a study of inhibitors against external corrosion.

Even with "airtight" packing, improperly packed tinplate can be rendered unfit for use by fretting due to movement of individual sheets, and stacks must be tightly packed and bound to prevent this from happening. If conditions are particularly hazardous, or there is a requirement for very high surface quality on delivery, components of cans may be lacquered externally by either roller coating in the sheet or spraying after forming.

A wide range of tests has been adopted to assess the susceptibility of a new system to atmospheric corrosions; they can range from simple steaming in enclosed boxes to very sophisticated automatically controlled cabinets operating to given cycles of humidity and temperature levels in an attempt to simulate tropical storage and sea journeys. These are sometimes extended to include injection of dust, SO_2 and salt sprays. All are essentially comparative, against a known control; and as translation of findings to performance in practice is very difficult, direct practical trials are usually required for final assessment of a new system.

REFERENCES

1. T.P. HOAR, *J. Electrodep. Tech. Soc.* **14,** (1983), 33.
2. D.R. GABE, *Principles of Metal Surface Treatment and Protection,* Vol. 28, Pergamon Press, Oxford, (1978), 193.
3. G.G. KAMM and A.R. WILLEY, *Proc. 1st Int. Cong. Met. Corr.*, 1961, 493.
4. A.R. WILLEY, *Brit. Corr. J.* **7,** (1972), 29.
5. J.C. SHERLOCK and S.C. BRITTON, *Ibid.,* **7,** (1972), 180.
6. S.C. BRITTON, *Tin versus Corrosion,* International Tin Research Institute, 1977.
7. J.C. SCULLY, *The Fundamentals of Corrosion,* 2nd Ed., Pergamon, Oxford, 1975.
8. G.W. PATRICK, *Anti-corrosion,* June 1976, 10.
9. E.F. KOHMAN and N.H. SANBORN, *Ind. Eng. Chem.,* **20,** (1928), 76.
10. L.H. JEFFEREYS, *Trans. Farad. Soc.* **20,** (1924), 392.
11. D.R. GABE and R.J. MORT, *J. Iron & Steel Inst.,* **203,** (1965), 64.
12. S.C. BRITTON and K. BRIGHT, *Corrosion,* **17,** (1961), 98.
13. R.A. COVERT and H.H. UHLIG, *J. Electrochem. Soc.,* **104,** (9), (1957), 537.
14. L.R. BEARD *et al, Trans. Inst. Met. Fin.,* **44,** (1966), 1; **49,** (1971), 63.
15. A.J. DISCOMBE *et al, J. Iron & Steel Inst.,* **203,** (1965), 1252.
16. C.C. KAMM *et al, Corrosion,* **17,** (1961), 84.
17. W.E. HOARE *et al, The Technology of Tinplate,* Arnold, London, 1965.
18. V.W. VAURIO *et al,* Determining the Corrosion Resistance of Tinplate, *Ind. Eng. Chem.,* **10,** (1938), 368.
19. R.R. HARTWELL, Corrosion Resistance of Tinplate: Influence of Steel Base Composition on Service Life of Tinplate Containers, National Metals Congress, Cleveland, 1940, American Society for Metals, Preprint 44.
20. R.P. FRANKENTHAL *et al,* Corrosion in Food Containers: the Mechanism of Corrosion of Tinplate by Various Food Products, *J. Agric. Food Chem.,* **7,** (1959), 441.
21. J.C. CONNELL and R.D. MCKIRAHAN, *Food Tech.,* **13,** (1959), 228.
22. A.R. WILLEY, J.L. KRICKL and R.R. HARTWELL, *Corrosion,* **2,** (1956), 433.
23. G.D. HARDEN and J.F. WALPOLE, The Influence of the Alloy Layer on the Performance of Lacquered Tinplate for Food Cans, Second International Tinplate Conference, ITRI, London, 1980.
24. J.J. DALY, Can Corrosion Problems, *Corrosion,* (National Assoc. of Corr. Engrs., USA), **15,** (11), Nov 1959.

25. E.L. KOEHLER, J.J. DALY, H.T. FRANCIS and H.T. JOHNSON, Corrosion Processes in Carbonated Beverage Cans, *Corrosion,* **15,** Sept 1959.
26. R.E. BEESE, A.R. WILLEY and G.G. KAMM, Canned Soft Drinks: a Study of Corrosion Mechanisms, 8th Annual Meeting Proceedings of Soft Drinks Technologists, USA, 1961.
27. M.G. ALDERSON, Packaging Problems of Carbonated Soft Drinks and Fruit Juices, *Food Mnfr.,* Aug 1970, 67.
28. M. MAHADEVIAH *et al,* Corrosion of Tinplate by Citrus Juices, *J.Fd. Technol.,* **11,** (1976), 273.
29. A. SEMEL and M. SAGUY, Effects of Electrolytic Tinplates and Pack Variables on the Shelf Life Canned Pure Citrus Fruit Juices, *J. Fd. Technol.,* **9,** (1974), 459.
30. M. LANDAU and C.H. MANNHEIM, *J. Fd. Technol.,* **5,** (1970), 417.
31. S.J. DULY, Condensation on Board Ship, *J. Roy. Soc. Arts,* **86,** (1938), 439.
32. W.F. SEYER and P.J. DONALD, Corrosion of Milk Tins in Ocean Transit, *Canad. Chem. Metall.,* **20,** (1936), 227.
33. R.K. COHEN and M. NELSON, Evaluation and Control of Sweat Damage, Stanford Research Inst. Report, Project No. S-2179.
34. G. BARBIERE *et al,* Inhibition of the External Corrosion of Tinplate Cans, *Industria Conserve,* **50,** (1975), 203-208.

The following papers in the *Proceedings of the Second International Tinplate Conference,* 1980:

35. ANA ALBU-YARON and D.A. SMITH, Application of Transmission Electron Microscopy to the case of Structural Surface Studies of Lacquered Tinplate.
36. S. HARADA *et al,* The Effect of Surface-enriched Elements in Killed Steel on Tinplate Corrosion Resistance.
37. M.E. WARWICK, Laboratory Studies of the Corrosion of Side-seams in Soldered Tinplate Containers.
38. M. TSURUMARU, Evaluation on Iron Exposure of Tinplate and in Tinplate Cans.
39. P.R. CARTER, Effect of Alloy Weight on Corrosion Performance, ATC Values and other Special Properties of Tinplate.
40. R. CATALA *et al,* Inhibition of Tin Dossolution in Canned Vegetables.
41. J. HAGGMAN, The Role of Hydrogen in the Corrosion of Tinplate Cans.

Chapter 8

Waste Recovery and Recycling

8.1 INTRODUCTION

The need for reducing waste to a minimum and effective recovery of scrap materials has of course been recognised for a long time, but development of more effective methods have received substantially increasing attention over recent years in almost all industrialised countries. Three major reasons account for the considerable effort now being given to improving recovery processes:

(i) An increasing need for saving valuable resources, particularly where known deposits appear likely to be exhausted in the fairly near future, or where a real reduction in the import bill is of paramount importance, or when they are generally available only from politically vulnerable areas.

(ii) A special need to save in cases where appreciable energy requirements are involved in the extraction process.

(iii) A compelling need to reduce the burden of conversion costs in an increasingly competitive environment.

In parallel with these economic requirements, increasing pressure has also been exerted for substantial reduction in the amounts of objectionable vapours and solid materials being discharged into the atmosphere, elimination of insanitary methods of refuse dumping, and control of dumping of poisonous wastes. Demanding new legislation has been enacted in many countries to achieve major reduction in pollution by industrial effluents. The particular problem of stoving oven effluent has been discussed in Chapter 6, Protective and Decorative Coating Systems.

Many improved methods of recycling are available or being developed, but development costs can be heavy, and considerable duplication of effort could occur, particularly in complex situations. Economics play a vital part in consideration of viability of recycling schemes. In general, they should neither use more energy nor cost more than the saving provided by recycling. If realistic assessment of any schemes indicates that they are excessive, political judgement for continuation, so as to avoid unacceptable environmental nuisance, must be accompanied by recognition that an essential service is being provided.

Recycling involves many technical and organisational difficulties, and persuasion must play a major role in achieving the co-operation needed between industry, local authority and users, in addition to national government departments. The need for effective control of effort and technical and financial assistance has led to the setting up of governmental agencies or sponsored assocations in many areas including the United States, the Far East and in Europe. A selected list of associations, etc., is given after the references at the end of this chapter.

In the United Kingdom, the Industry Committee for Packaging and the Environment (INCPEN) was set up some years ago, amongst other aims, "to look for ways to stimulate and assist its members (packaging industry, fillers, distributors and retailers) to implement and expand recycling schemes". Their literature, dealing with conserving raw materials and energy, reduction in pollution, economic benefits and practical difficulties involved, are generally available.[3] Similar help is available from similar organisations in most industrialised countries. Many other aspects frequently need to be taken into account, e.g. public health matters, but these are outside the scope of this discussion.

In parallel with these activities, demands are increasing under consumer protection legislation for reduction in what have become known as "heavy metals" in fresh foods and all forms of packaged food, particularly in metal cans. The main elements in the case of food cans are lead and tin, but others such as antimony, arsenic, bismuth, cadmium and mercury can be involved in some forms of packaging. Most countries have restrictive or advisory agencies to limit the potentially dangerous constituents in all fresh foods and drinks, and the pick-up that can take place from packages. The well-known Food and Drug Administration (FDA) in the United States and the British Industrial Biological Research Association (BIBRA) are just two examples. The FDA is a federal agency, and its code of regulations are enforceable under US law; BIBRA is an industrial assocation formed to provide advice to their members on biological matters, but their guidance carries some weight under British law.

The latest Advance Notice of proposed rulemaking on Lead in Food was issued on 31 August 1979 by the US Dept. of Health,

Education and Welfare Food and Drug Administration in *Federal Register*, Vol. 44, No. 171, 51233; an editorial summary of the FDA Lead Reduction Programme was given in the *Almanac of the Canning, Freezing, Preserving Industries*, 68th edition, updated to 15th May 1983.

Some further information on the incidence of "heavy metals" in the environment and in food is given under Further Reading, after the references at the end of this chapter.

Allied to these general activities are those particularly geared to foods and articles consumed by or accessible to children. Childproof closures have been developed for containers of pharmaceutical and other products. Many printing inks used to contain significant amounts of lead oxide and/or other toxic metal compounds, but as a result of the activities of many organisations, these hazards are well on the way to being completely eliminated on a worldwide basis.

TABLE 8.1.1. RESOURCES USED AS RAW MATERIALS FOR ALL PACKAGING CONSUMED IN THE UK

1979/80, thousand tonnes

	METAL	PAPER	GLASS	PLASTICS	TOTAL
Iron ores & scrap	1,740	–	–	–	1,740
Limestone & dolomite	130	–	800	–	930
Salt	15	–	660	–	675
Bauxite	250	–	–	–	250
Sand & gravel	–	–	1,200	–	1,200
Pulp & fibre	–	2,593	–	–	2,593[a]
Oil	45[b]	–	–	1,520[c]	1,565
Coal (non-fuel use)	470[d]	–	–	–	470
Tin	5	–	–	–	5
Other materials	340	105	60	95	600

Some indication of the scale of use of materials for packaging can be gained from Table 8.1.1, and it must be appreciated that these amounts are for only 55 million people, out of at least 1000 million in the larger industrialised countries who make rather similar use, on average, of packaged foods. Tinplate packaging in the United Kingdom has been examined in more detail by Thomson,[2] (Fig. 8.1.1 (a) and (b)) and the same observations can be made as to the wider implications. Data of this type can only be estimates, as they must be based on sampling schemes and the use of conversion factors; they must not, therefore, be regarded as absolute quantities.

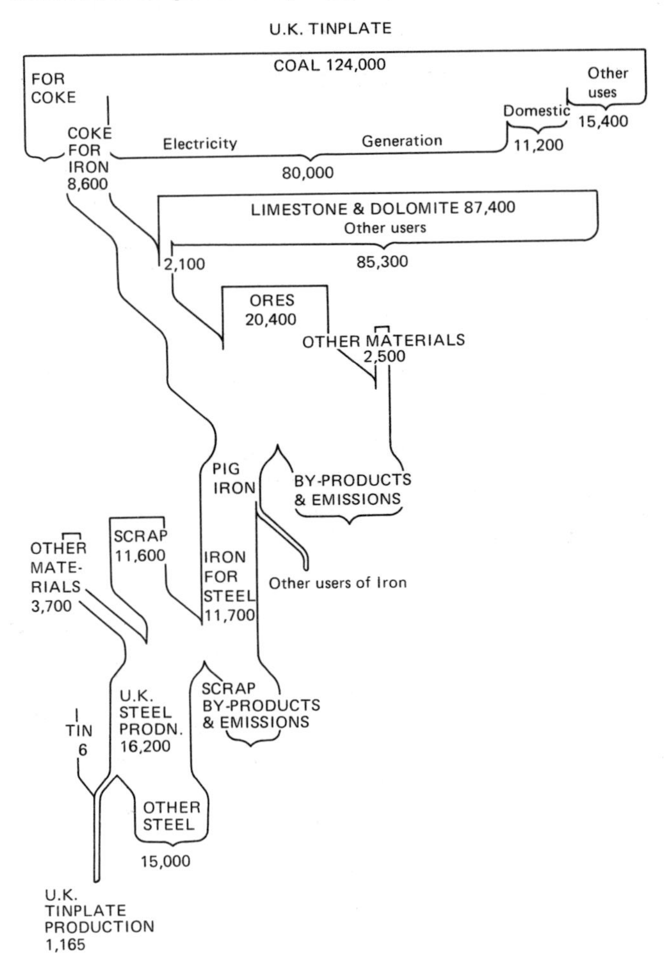

FIGURE 8.1.1a UK TINPLATE RESOURCE USE (THOUSAND TONNES, 1977)

In parallel with these means of material saving and recovery, continued effort has been applied to reducing the metal content of cans by the use of thinner, stronger, plate and by incorporating strengthening beads into can walls and other forms of redesign. An often quoted example of material saving the United Kingdom is that for a 12 oz beverage can, the metal content of which has been reduced by 40% since 1970, from 57.5 g to 36 g.

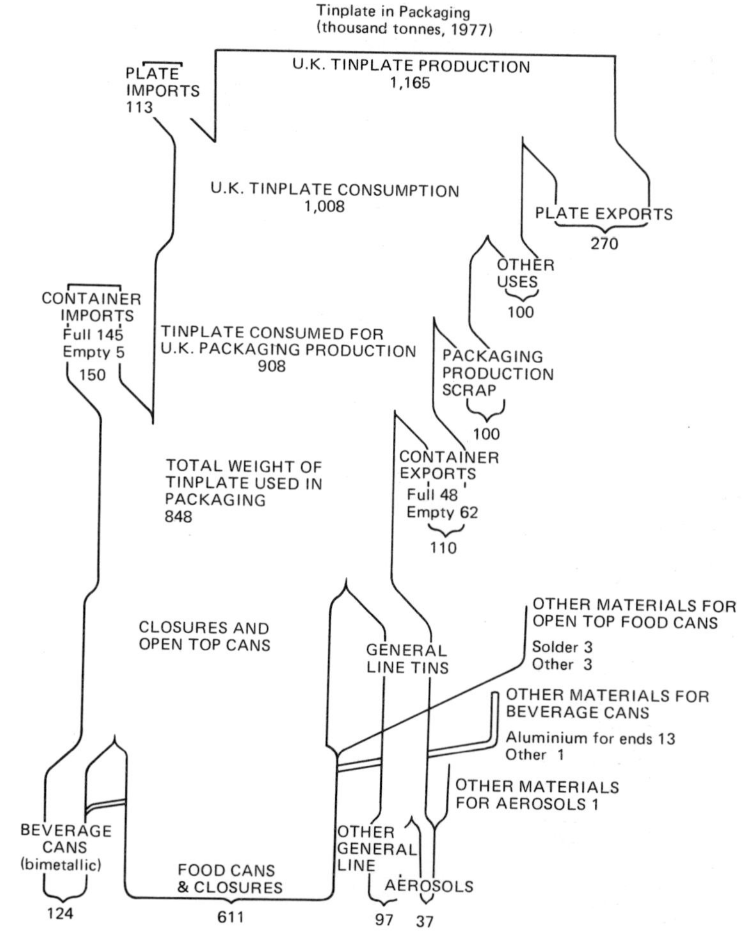

FIGURE 8.1.1b TINPLATE IN PACKAGING ESTIMATES (THOUSAND TONNES, 1977)

8.2 RECYCLING OF INDUSTRIAL TINPLATE SCRAP

The recycling of industrial tinplate waste ("clean" tinplate scrap) has been practised for a long time. Goldschmidt developed a detinning process in Germany in 1885, Batchelor's operations in England started in 1889, and the Vulcan enterprise in the United States in 1902. Detinning has been practised in most parts of the world for a number of years.

Clean tinplate scrap arises to some extent during tinplate manufacture, but especially in canmaking. Overall, some 15% inevitably ends up as waste, off-cuts and spoiled work. As up to nearly 70% of total canmaking costs arises from the cost of tinplate, effective detinning has provided a valuable saving. Currently in the United Kingdom, industrial tinplate scrap amounts to about 150 000 tonnes per annum; apart from some 15 000 tonnes fed into ironmaking furnaces. The bulk of this is subjected to detinning, thus providing good quality steel scrap containing less than 0.03% residual tin — British specification calls for less than 0.05% tin for steelmaking — and a good return of high purity tin. An account of UK practices was given by Linley.[1]

An electrolytic detinning process was used for some years in the United Kingdom, but this has now been discontinued. It consisted of treating small quantities (75-150 kg) of scrap held in steel mesh baskets and immersed in hot caustic soda (or potash) solution; to obtain reasonably high rates of deposition the solution was maintained at near boiling point. Tin collected on the cathodes in sponge form was periodically scraped off, and melted and cast; about 1 kg of tin was deposited per 1000 A-hr. Any lacquer and print present on the plate surface had to be removed prior to electrodeposition, either thermal or chemical methods being used. The stages involved in the whole process are illustrated in Fig. 8.2.1.

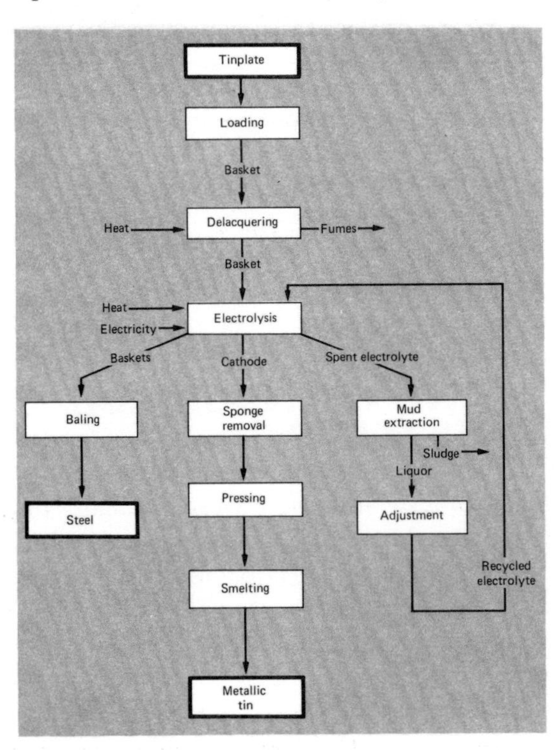

FIGURE 8.2.1 PROCESS DIAGRAM — ELECTROLYTIC DETINNING

Good quality steel scrap is produced by the process, and, after washing, it is baled to a specified density for sale to steelmakers. Tin yields also are good.

This is by its nature a batch process, in the most highly automated versions of which the scrap is handled several times. This process is practised in many European works, the largest of which in Germany has a capacity of 30 000 tonnes per annum.

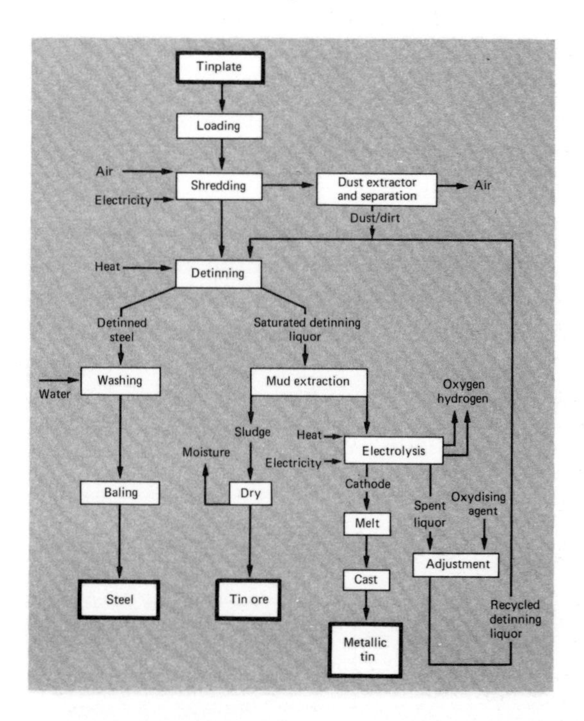

FIGURE 8.2.2 PROCESS DIAGRAM — ALKALINE DETINNING

Alkaline detinning

The rate of detinning in alkaline solutions is very slow, but it can be considerably increased in the presence of oxidising agents. This process is used by Bachelor Robinson (now Vulcan Materials) in the U.K., and is capable of removing the tin from electrolytic tinplate in 4 hours in a static bath, which is reduced to 1½ hours in the automated continuous process developed by them. The free tin is removed in a few minutes, the additional time being required to remove the tin/iron alloy. Two plants operating the process (its details are shown in Fig. 8.2.2) handle together some 135 000 tonnes of tinplate scrap per annum. This covers a wide range of form (from whole sheets to rejected cans) and the feed has to be disintegrated to give pieces roughly 2.5-15 cm maximum dimension by means of high-speed hammer mills. The bulk density of the material will then be between 320 and 400 kg/m^3; it is conveyed through the detinning tanks with a residence time of about 1½ hours, the tinplate and solution being kept continuously in relative movement. The alkaline tin solution is run off, and after removing suspended solid by centrifuging it is electrolysed and the tin recovered by melting it off the cathodes and refining, giving a final purity of over 99.9%; the solids contain significant amounts of tin and this also is recovered. Overall yield is about 88%.

8.3 RECYCLING OF USED CANS

From Fig. 8.2.3[1] it can be seen that 1.1 million tonnes of tinplate were consumed in the United Kingdom in 1975. About 150 000 tonnes of industrial waste is generated in the manufacture of tinplate goods, mainly cans, and the processing of this for recovery has been described. On the balance; about 85%, equivalent to 800 000 tonnes of tinplate, or 795 000 tonnes of steel and 5000 tonnes of tin, is lost, mainly by burying with domestic refuse.

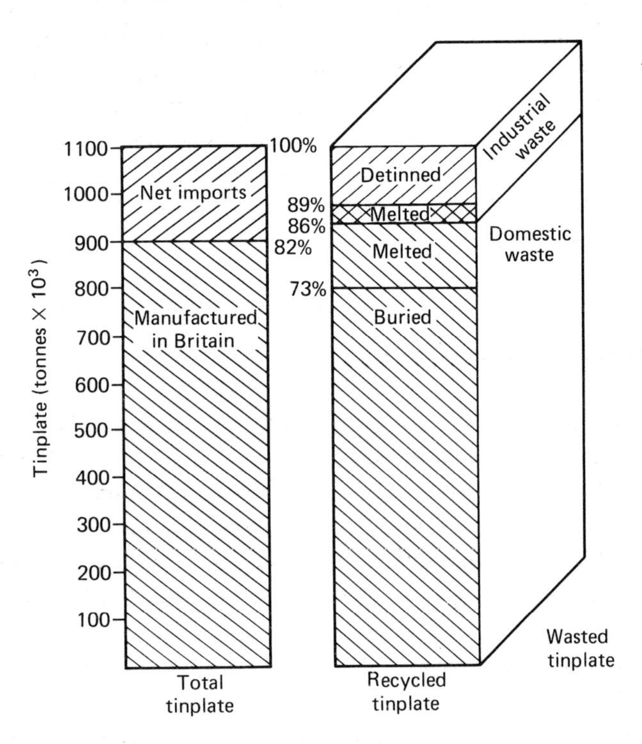

FIGURE 8.2.3 TINPLATE MANUFACTURE AND DISPOSAL

Many attempts have been made in the past to extract the cans, and between 1939 and 1945 considerable quantities of tin and steel were recovered. From 1945 until relatively recently the operation was carried out only on a small scale because of disinterest on the part of central and local authorities and the small margin of profit resulting from a complex process. This picture has been changed by:

1. The setting up by the U.K. Government of the Industry Committee for Packaging and the Environment with the object of advising and encouraging manufacturers and users of all packaging materials, coupled with local authorities, to develop effective means for recycling.
2. The joint activity of Metal Box, British Steel Corporation and Batchelor Robinson (now Vulcan Materials) through the setting up of Material Recovery Ltd. in 1975 to promote a study of the technical and economic feasibility of extracting used cans from domestic refuse and its recycling, and develop techniques, in collaboration with many local authorities.
3. Schemes promoted by Can Makers (a new grouping of five major U.K. can manufacturers formed in 1981) and their major suppliers (British Steel Corporation and Alcoa) to expand substantially methods of used can recovery. These include: Material Recovery Ltd., promoting further reclamation units in conjunction with additional local authorities, coupled with providing an engineering consultancy service, and improved methods for separating aluminium ends in bimetallic cans. It is anticipated that recovery of used cans, tinplate and aluminium, by the three schemes in operation, will be increased by over 50% by 1985. Used cans deposited at collection sites are much to be preferred, as they are considerably cleaner than those extracted from domestic refuse.

These increased activities have also been encouraged by the EEC throughout European countries, and indeed worldwide. The U.S. Government and Australia also set up agencies some years ago to encourage development of improved and wider recovery systems. In the case of aluminium cans, Alcoa estimate that in Australia over 50% are being recycled, with a level approaching 50% in the United States. Increased effort has also been applied in most major countries to developing improved preparation and detinning technology.

There are three important stages in the recovery of cans from refuse:

1. *Extraction:* As Fig. 8.3.1. shows, over 90% of other contaminating materials have to be rejected.

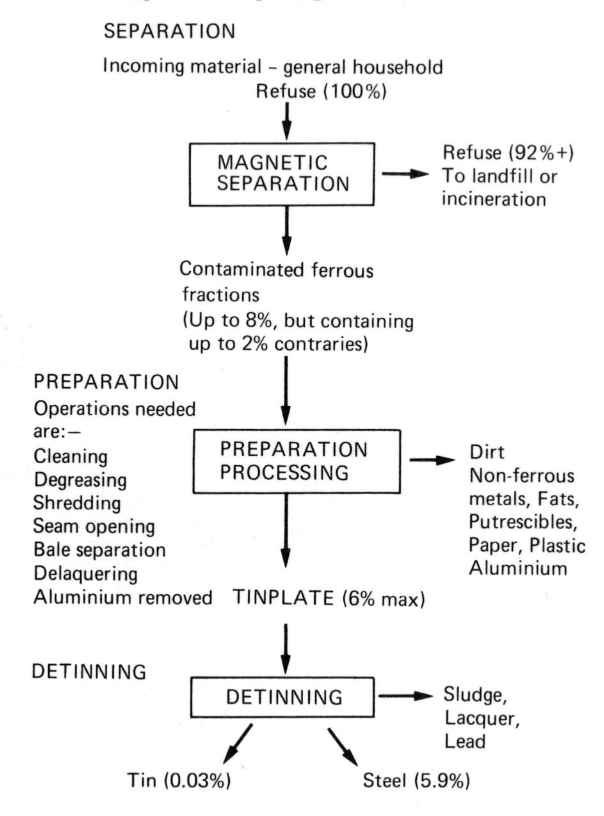

FIGURE 8.3.1 PROCESS DIAGRAM — USED CAN RECYCLING

2. *Preparation:* This involves cleaning the recovered cans free of decayed matter and other contaminants, and converting them into a physical state similar to that of industrial tinplate waste, to be suitable for detinning.
3. *Detinning:* Which, as described earlier, includes purifying the recovered tin and upgrading the residual steel scrap. The two widely used detinning methods are described in sections 8.2.1 and 2.

The preparation process is illustrated in Fig. 8.3.2. Several methods can be used for separating tinplate cans from the general refuse: a typical system consists of decompacting the refuse, magnetic separation of all steel constituents, extracting heavy steel articles, shredding residual steel articles (almost entirely cans) and separating out their dirt content.

Shredding the cans and cleaning them free from dirt is a vital

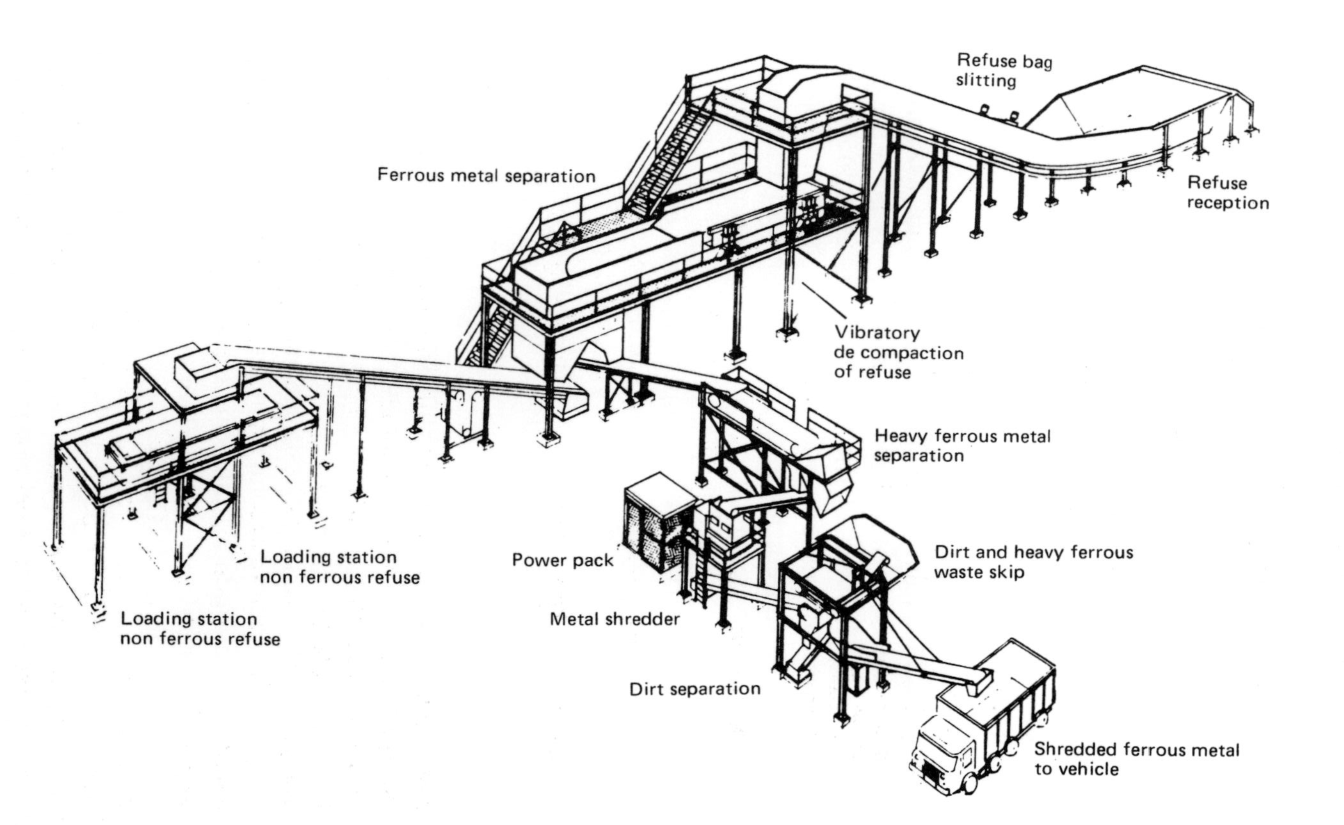

FIGURE 8.3.2 MATERIAL RECOVERY LIMITED, STALYBRIDGE PLANT

stage if effective detinning is to be technically successful. A specification developed by Batchelor Robinson & Co. has the following requirements:

(i) All seams opened to at least 90% of their total linear dimension (up to 29% of the tin content can be within the seams), to allow access of the detinning solution.
(ii) Dirt content must be less than 1% (to minimise fouling of the detinning solution).
(iii) Residual aluminium content less than 0.2% (reacts with caustic contents).
(iv) Shredded pieces should be between 2.5 cm and 15 cm maximum.

In addition, its bulky density should be greater than 400 kg/m^3.

Detailed descriptions of the first Metal Recovery Plant are given by Linley[1] and Thomson.[2] The latter also discusses in detail the economic aspects of resource recovery, highlighting several constraints (capital required; ferrous content of and total amount of refuse available for treatment; its tin content; stringent material specifications; etc.). Some examples are given of capital and operating costs. The ASTM (USA) specifications quoted include chemical and physical requirements of the tinplate scrap; for the recovered ferrous scrap, for foundry use and for iron and steel making; together with maximum Al and minimum Sn contents of the detinners raw material. The latter is becoming more difficult to meet, as the trend in many areas towards lower average tincoating weights is continuing; paradoxically, lower tin contents assist disposal of undetinned recovered tinplate scrap in cast iron and iron and steelmaking operations. The beneficial effect of higher tin prices for the detinning processes but the low and variable value of steel scrap are also dealt with.

Thomson refers to alternative fragmenting methods under development, and describes a second plant set up in co-operation with a local authority to produce some 9000 tonnes of used tinplate scrap per annum. A recent review of current activities and further action required is given in the INCPEN Recycling Report, 1982.[3]

8.4 ALTERNATIVE RECOVERY METHODS

Although the recovery of cleaned, used, tinplate scrap, followed by detinning, is usually the preferred method, several alternatives are sometimes used:

(i) As a source of cheaper ferrous scrap in the manufacture of lower grade iron castings, and in "mini" steel mills; as described in section 8.2, the tin is lost in both cases.
(ii) As a source of tin for "refined" iron for good wearing properties; this use and examples were also described in section 8.2.

The amounts used in these applications appear to be decreasing.

Much of the residual material (paper and plastic) is combustible and can be regarded as a useful fuel; it is said to have a calorific value of about 10.3 MJ/kg[4] (roughly half that of coal), but will vary according to moisture content.

The largest incinerator in the United Kingdom (Edmonton, London) was installed in 1970, and is used for the generation of electricity; as its capital and operating costs proved to be higher than anticipated, further large-scale installations have not been built. This unit, however, is operating consistently, and in fact is London's cheapest modern method of disposal; it is claimed that if the steam generated was sold directly to industry, disposal costs would be even lower. Three other incinerators are in use supplying steam to local organisations or for district heating.

FIGURE 8.4.1 PELLETISED WASTE DERIVED FUEL

One U.K. industrial organisation uses shredded waste, after removal of the ferrous fraction, as supplementary fuel to coal in stoker-fired boilers for internal electricity generation. Another supplies it as a supplementary fuel in cement kiln operation, the ash being incorporated as a constituent in their cement mixtures.

A further type is the formation of pelletised waste derived fuel (WDF) which can be used in long-term storage; this process is employed in three plants within the United Kingdom. Figure 8.4.1 illustrates the fuel pellets produced. Its calorific value is usually in the range 10-20 MJ/kg, generally a little more than half that of coal, depending again on moisture and plastics contents; the material's slightly higher ash content is not a major problem in modern boiler furnaces.

Many other systems are possible, and several design projects are in hand. A basic requirement of these systems is that up to 200 000 tonnes of waste per annum need to be available at one site.

REFERENCES

1. B.D. LINLEY, Recycling of Tin and Tinplate, First International Tinplate Conference, ITRI, London, 1976.
2. J.F. THOMSON, The Recovery of High Grade Steel Scrap and Tin Scrap from Refuse-derived Used Cans. Second International Tinplate Conference, ITRI, London, 1980.
3. *Recycling*. The Industry Committee for Packaging and the Environment, London, 1982.
4. *The Doncaster Project*. Warren Spring Laboratory.
5. Report 1974 – Recycling Investigations, Private Report Metal Box plc; British Steel Corporation Batchelor Robinson plc, London, 1974.
6. American Society for Testing and Materials (ASTM). E701 Standard Methods of Testing Municipal Ferrous Scrap. E702 Standard Specification for Municipal Scrap. ASTM, Philadelphia Pa., USA.

FURTHER READING

HARVEY ALTER, Development of Specifications for Recycled Products, First World Recycling Congress, Basle, 6-9 March 1978. Reprinted in *Conservation and Recycling*, Vol. 2, pp. 71-84, Pergamon Press, 1978.

E. JOSEPH DUCKETT, The influence of Tin Content on the Re-use of Magnetic Metals Recovered from Municipal Solid Waste, *Resource Recovery and Conservation*, No. 2 (1976-77), 301-328.

T.D. NEILSON and C.I. BATES, Pulling Iron out of the Fire, Paper by Material Recovery Ltd., Reading, 1980.

Organisations Concerned with the Development of Recovery Systems and Techniques (additional to those mentioned under references)

USA

National Centre for Resource Recovery, Washington, D.C.
U.S. Department of the Interior, Bureau of Mines, Washington, D.C.
National Association for Recycling Industries
Committee of Tin Mill Product Producers, AISI, Washington, D.C.
Institute of Scrap Iron & Steel, Washington, D.C.

together with major steelmakers, can manufacturers.

United Kingdom

U.K. Department of the Environment, Warren Spring Laboratory, Stevenage, Herts. (development of new techniques)

Netherlands

Central Technisch Institute (TNO)
The Institute of Waste Disposal Working Party – Re-use of Tinplate

Switzerland

WNO – International Reference Centre for Waste Disposals, Dubendorf

West Germany

Recycling Weissblechpackungen (Recycling of Tinplate Packages), (attached to the Informations-Zentrum Weissplach, Dusseldorf)
Umweltbundesamt (the authority for ecology), 1000 Berlin, 33 Postfach

Japan

The Agency of Industrial Science and Technology, Tokyo

Australia

Steel Can Group (consisting of representatives from the Broken Hill Proprietory Co., several container manufacturers and others interested in steel can recycling)

and similar organisations in most industrialised countries.

Lead and Health, (Report of a Working Party of Lead in the Environment), Department of Health and Social Security, HMSO, London, 1980.

Heavy Metals in Foods, British Food Manufacturing Industries, (BFMIRA), Leatherhead, England, 1980.

Pollution in the Atmosphere, Final Report on Study Group, The Royal Society, London, 1978.

Food Safety, University of Texas Health Science Centre, 1980, Final Report.

Lead in the Human Environment, National Academy of Sciences, Washington, 1980.

Survey of Lead in Food, (Working Party on the Monitoring of Foodstuffs for Heavy Metals), Ministry of Agriculture, Fisheries and Food, HSMO, London, 1982.

Recent European Legislation on Foods, First Annual Lecture, Campden Food Preservation Association, Campdcn, England, 1979.

Lead in Food Regulations, HMSO, London, 1979.

List of Maximum Levels Recommended for Contaminants, the Joint FAO/WHO, Rome, FAO/WHO, 1978.

Lead in Canned Foods, Comite Interprofessional de la Conserve Medecine et Nutrition, France, 1981.

Harmonisation of Legislation on Foodstuffs, Food Additives and Contaminants in the European Economic Community, II, *Journal of Food Technology*, **13,** (6), (1978), 491.

Sources of Lead Pollution – a Review, *British Medical Journal,* **282,** (6257), (1981), 41-44.

Chapter 9

A View of Future Developments

Remarkable progress has been made in developing, to a fully commercial stage, two new high-speed sophisticated processes — the Draw and Wall-ironing and the Draw and Redrawn — within a space of only 15 years. Undoubtedly one of the major reasons for this success has been the very close co-operation which has been built up between all raw material manufacturers and canmakers, to an extent which would not have been thought possible 30 years ago.

The growth potential for packaged foods and beverages is still very high; *per capita* consumption in most industrialised countries is still much below that in the United States, and many are very low in comparison. When one considers some projected world population figures — some 7000 million people by the year 2000[1] — tremendous food requirements will need to be met in areas far from major growth regions. It is said that half the food grown by many developing countries does not reach their own underfed people, because of inadequate packaging and transport facilities. There is no doubt that the potential for considerable growth exists, but competition from other major packaging — especially glass and plastics — has become very keen, and these are not likely to relax their extensive efforts towards obtaining more of the business in the future. It appears that more major cost-cutting developments are taking place in aluminium production methods than in steel or glass; although the marketing policy of suppliers dictates prices, this is mainly in the short term. Costs rising at different rates are bound to upset price competition, but ultimately, the cost per 250 ml or per 500 g of a product will be the major customer criterion.

In addition to the urgent need for many major political decisions, the basic requirements for further improvement to packaging are:

Improved pack integrity, especially under increasingly adverse handling conditions.

Reasonable prices (more difficult in times of recession).
Good customer appeal.

What needs to be done in the case of low carbon steel sheet and its conversion into containers, is discussed broadly in the following pages. In summary, these amount to continuing improved fabrication efficiency and material quality all along the line, lower scrap levels and labour content, coupled with stronger material, and increased strength by beading and other techniques. Whether these measures in themselves will be adequate, however, will always be debatable.

This view of the future is not intended to be a detailed prediction; extrapolation is fraught with considerable danger, and there is much relative to materials and forming that is still not known. However, the future will require considerable effort and co-operation, and undoubtedly a need for regular reassessment.

Cheaper packages will basically involve employing thinner material, and in this respect low carbon steel seems to have the edge on most other materials, especially when bearing in mind the more hazardous transport conditions usually found in less developed countries. At this stage there is no shortage of tinplate, and many three-piece can lines can be fairly readily converted to side seam welding.

9.1 IMPROVED MATERIALS

The extensive co-operative work carried out by steel manufacturers and canmakers has emphasised the basic needs for continuous improvement: (i) cleaner steel, (ii) more uniform geometrical characteristics, and (iii) more uniform metallographic structures; efforts to improve these further, and to achieve greater uniformity, will continue. This is vital to enable more severe forming operations to be developed, and to allow lower thicknesses to be used.

As operational speeds have become too fast for normal human reaction to detect and deal with breakdown quickly enough, better control of all processing factors will call for more precise electrical or electronic methods coupled to computer controlled linked systems; these will include all rolling operations and electrotinning, coupled with more efficient loading and faster operational speeds. Truly continuous rolling has already been achieved in several plants, and general improvement on a broad front is widespread.

In aiming for cleaner steel, especially to avoid the incidence of the larger, harmful NMIs, use of secondary steelmaking and continuous (strand) casting will grow; but being capital intensive, large investments will be needed, and wider adoption is bound to be gradual.

Extensive research being applied to improving further the hot- and cold-rolling processes will continue to improve efficiency and strip quality (uniformity of gauge and flatness), but again the rate of improvement will closely related to economic conditions and the likely demand. In view of the substantial improvements made in gauge and tincoating mass uniformity, there seems to be a strong case for agreement to tighter basic standards, at no extra cost.

Considerable publicity has been given to the savings possible by reduction in tincoating weight used and by the use of alternative metallic or chemical surface treatments. Many of these can be misleading, e.g. direct comparison of the coating metal cost per unit ignores other important factors. The beneficial effect of tin on ironing in the DWI process has been clearly demonstrated, and a real increase in ironing load could have many serious effects, such as greater tool wear, more down-time, the need for more efficient lubrication systems, and even better lacquer application. As the beneficial effect of tin applies mainly to the outside surface, i.e. that in contact with the ironing rings, one of the first reductions made was of the internal coating weight down to 1.00 g/m^2, as part of a specification of 2.50/1.00 which provides a useful saving. This trend toward lower tincoating weights, which are non-standard grades, will continue to the limit. All will require very effective lubrication systems and the final container will need to be thoroughly tested against the intended content. The current high cost of tin, over £9000 per tonne in the UK, has led to numerous tests on the possibility of using untreated blackplate; the apparent significant saving in material cost may, however, be reduced by:

(i) Any surface treatment that may be required (either of the steel surface initially or after wall-ironing).
(ii) More costly lubricant systems.
(iii) Special tooling (ceramics have been proposed) and more frequent tool changes.
(iv) The possiblity of additional or special lacquering treatment.

Some authorities are of the opinion that untreated blackplate will never be wholly satisfactory, but that more effective chromate/phosphate type treatments will be developed which will, coupled with improved lubricants and/or lacquer systems, provide an adequate performance. Major changes of this type will involve extensive commercial testing. Chrome/chrome oxide materials cannot be used in the DWI process in view of the abrasive nature of its surface. Newly developed electroplated nickel coatings, at a weight of about 0.20 g/m^2 (over tenth that of typical tincoating) are also under active testing for the DWI process.

In view of the considerable difference in price between tinplate and tinplate scrap, many times more than is the case with aluminium, considerable attention is also being given to reducing the scrap levels which can occur in a DWI type process. Good toolmaking practice, precise toolsetting and maintenance schemes can also contribute to possible savings.

The DRD process does not impose the same restriction on tincoating reduction, and in fact chromium/chromium oxide coated steel sheet is already being used commercially for products where the anodic protection provided by tin is not essential; lacquer adhesion, a vital requirement in DRD cans, is generally equal to that on tinplate. In view of the greater metal usage involved in producing DRD cans, attempts are also being made to use cheaper chromate/phosphate treated blackplate.

Attempts are also in hand to reduce plate cost by using intrinsically stronger steel, and therefore thinner plate. Some DWI manufacturers are already using the CA temper T61 ("T4 CA"), which has allowed its thickness to be reduced to 0.29 mm. Extensive trials are being carried out on DR plate at a thickness of 0.27 mm.

Trends towards the use of stronger plate have gone further in the case of the DRD process, where two grades of DR plate — DR 550/620 ("DR8/DR9" BA) and DR620 ("DR9", CA) — are already used commercially in some areas.

In both cases, considerable improvement to the process has been required, with extended commercial trials to ensure that these changes are acceptable. Application of both can types is being widened to include increasingly pet food, meat and vegetables products; as these are steam processed, some strengthening is required, usually by introducing beads into the can wall and modification of the base profile.

Extensive trials were carried out in the sixties and seventies to increase substantially the strength of low carbon steel; these have been described by Gibbon,[2] Jefford,[3] Berwick[4] and McFarland.[5] By quenching the steel at very high cooling rates from subcritical, intercritical and supercritical annealing temperatures, a range of strengths much higher than those of the strongest DR plate were achieved, but none was commercially viable largely because of severe shape problems, surface oxidation and critical dependence on carbon content. Jefford quotes strengths in the range 420-1600 $N.mm^2$ being achieved on quenching from 700° up to 900°C in a pilot line; the material was processed to finished tinplate and subjected to canmaking trials, but the process was judged to be not viable at that time. Berwick and McFarland describe the somewhat similar Inland Steel development carried out in a commercial strand annealer and using very rapid quenching rates; the material, termed "martensitic" plate, was available commercially for a time, but was then withdrawn.

An alternative to rapid quenching was an examination of the possibilities of partial annealing of cold-reduced strip, so as to retain part of the work-hardening induced during cold rolling. This was studied in detail at Youngstown Sheet & Tube, and is described by Toth;[(6)] in this case also, a continuous annealer was used. It was shown that material having a strength of 650 N/mm^2 could be produced, which was suitable for end stock; this was achieved by subjecting the plate to the recovery stage only (Chapter 2, Annealing). This process also was judged to be too critical, as steel composition in relation to time/temperature was very sensitive. Addition of alloying elements to the steel have also been attempted as means of increasing its strength.

In view of the substantial progress made in control systems for physical conditions (time, temperature, shape, etc.), it is arguable that these early attempts should be re-examined.

Another trend in DWI process development is the planning of so-called mini lines, for situations where the substantial capital cost of a full high-speed line would not be justified. These would consist broadly of:

(i) a "cupper" press, an "off the shelf" type, producing nearly 500 cans per minute;
(ii) two bodymakers, hydraulically operated, with high-speed trimmers; and
(iii) 500 cans per minute printers plus UV curving.

The investment required for this type is no greater, on a "per can" basis, than the massive high-speed lines.

Equipment cost for the DRD process is comparatively low, but it does not offer the same material saving as in the DWI process. Developments are therefore mainly in the direction of producing taller cans.

Height:diameter ratios greater than 1.0 are already being used, but there is a body of opinion that H:D ratios greater than 1.20 will not be possible, dependent on can diameter. Can diameters greater than 100 mm are not likely, either. These size dimensions, however, would substantially increase the number of products which could be packed economically.

Necking-in of the can orifice is likely to be developed in parallel, in view of material saving effect of smaller diameter ends.

The number of can sizes used commercially, and therefore the number of tinplate sheet sizes, is very high, causing many size changes in tinplate and canmaking operations. Reduction of the total number in use would offer substantial economy; this is an area which demands further attention jointly by packers and canmakers.

The wasted space consequent upon the use of circular-section cans has been commented on for a long time; it causes higher storage, packaging and transport costs than is necessary, and it has been calculated that savings of up to 30% could be made by adopting rectangular cans. Extra costs would have to be set against the apparent savings, due to greater manufacturing costs, coupled with more difficult manufacturing operations, to meet the product integrity needed. But there does seem to be a need for further consideration of these likely savings.

Many alternative forming processes are extensively used in other areas; these are mainly:

stretch forming,
swaging,
high energy and explosive forming processes.

It is generally argued that these are far too slow for the very rapid mass production methods needed for the packaging industry, but similar comments were made about welding some years ago. It would seem to be worthwhile to examine critically the possibilities of these alternatives in a rapidly changing world, especially in the light of further uses for tinplate. Other materials have made substantial inroads into the share of the market taken by the very long established tinplate food can; is there not a lesson to be learnt from these other successes, particularly in view of the close and wide ranging co-operation over a long period between plate manufacturers, canmakers and machinery manufacturers? Horizontal integration can often be more effective than vertical, given close understanding, critical analysis and above all real attention to detail.

REFERENCES

1. G. Habenicht, Tinplate Containers in a Changing World of Technology, First International Tinplate Conference, ITRI, London, 1976.
2. W.M. Gibbon, P.W. Davies and B. Wiltshire, *J. Iron & Steel Inst.*, **207**, (1967), 910.

3. G. JEFFORD, Steel Based Packaging Materials, *The Metallurgist & Materials Technologist*, Oct 1975, 516.
4. L.M. BERWICK and W.H. MCFARLAND, Paper to National Metal Congress, Chicago, 1966.
5. W.H. MCFARLAND, Paper to General Meeting AISI, New York, 1968.
6. R.G. TOTH and H.N. LANDER, Paper to General Meeting AISI, New York, 1969.

General Aspects

7. C.I. MELLOR, Some General Challenges to Tinplate Packaging, First International Tinplate Conference, ITRI, London, 1976.
8. D. GRETHER, Promoting Tinplate in Western Europe, Second International Tinplate Conference, ITRI, London, 1980.
9. C.I. MELLOR, The Friends of the Tin Can, Second International Tinplate Conference, ITRI, London, 1980.
10. D. WILLIAMSON, The impact of Tinplate Prices on the Packaging Industry, Second International Tinplate Conference, ITRI, London, 1980.

Index